INTERNATIONAL OVERVIEW ON SOLID WASTE MANAGEMENT

INTERNATIONAL OVERVIEW ON SOLID WASTE MANAGEMENT

A Report from the International Solid Wastes and Public Cleansing Association (ISWA)

EDITED BY

LUCA BONOMO
ISWA Scientific Secretary, Politecnico di Milano (Italy)

A. E. HIGGINSON M.B.E.
Past President, Institute of Waste Management

1988

ACADEMIC PRESS

Harcourt Brace Jovanovich, Publishers

London San Diego New York Berkeley Boston
Sydney Tokyo Toronto

ACADEMIC PRESS LIMITED
24/28 Oval Road
London NW1 7DX

United States Edition published by
ACADEMIC PRESS INC.
San Diego, CA 92101

British Library Cataloguing in Publication Data
is available

ISBN 0-12-114975-7

Typeset by Paston Press, Loddon, Norfolk and printed in Great Britain by
St Edmundsbury Press, Bury St Edmunds, Suffolk

ISWA National Members Contributing to the Present Volume

AUSTRIA OESTERREICHISCHER STADTEBUND, Rathaus, A 1082 Wien, Tel: 222 42 800.

CANADA AMERICAN PUBLIC WORKS ASSOCIATION (A.P.W.A.), Canadian Branch, 1313 East 60th Street, Chicago, IL. 60637, U.S.A., Tel: 312 667 22 00.

DENMARK DANSK KOMITE FOR AFFALD (DAKOFA), c/o Dansk Ingeniorforening, Vester Farimagsgade 29–31, DK-1606, Copenhagen V, Tel: 1 156565.

FEDERAL REPUBLIC OF GERMANY Verband Kommunaler Städtereinigungsbetriebe (V.K.S.), Lindenallee 13–17, Postfach 51.06.20, D 5000 Köln 51, Tel: 221 377 12 80.

FRANCE ASSOCIATION GENERALE DES HYGIENISTES ET TECHNICIENS MUNICIPAUX (A.G.H.T.M.), 9, rue de Phalsbourg, F 75854 Paris Cedex 17, Tel: 1 42 27 38 91.

HUNGARY ASSOCIATION OF PUBLIC CLEANSING AND OF SANITATION OF SETTLEMENTS, Rippl Ronai u.34–36, P.O.B. 197, H 1068 Budapest VI.

ITALY INTERNATIONAL SOLID WASTES AND PUBLIC CLEANSING ASSOCIATION (I.S.W.A.), Sezione Italiana, Viale Isonzo 14/1, 20135 Milano, Tel: 02 54 58 566.

JAPAN JAPAN WASTE MANAGEMENT ASSOCIATION, Kanda-Seiso-Jimusho, 1.1.6. Sotokanda, Chiyoda-ku, Tokyo 101, Tel: 3 251 40 68.

NETHERLANDS NEDERLANDSE VERENIGING VAN REINIGINGSDIRECTEUREN (N.V.R.D.), p/a Technische Dienst S.O.W., Postbus 201, 1620 AE Hoorn, Tel: 02290 41444.

NORWAY NORSK RENHOLDSVERKS-FORENING (NRF), c/o Oslo Renholdsverk, Gunnar Schjelderups vei 11, 0485 Oslo 4, Tel: 22 71 50.

POLAND POLISH ASSOCIATION OF SANITARY ENGINEERS AND TECHNICIANS, Czackiego 3/5, P 00–043 Warszawa, Tel: 22 26 28 94.

PORTUGAL ASSOCIACAO PORTUGUESA PARA ESTUDOS DE SANEAMENTO BASICO (A.P.E.S.B.), Rua Antero de Quental 44, 1100 Lisboa.

SPAIN ASOCIACION TECNICA PARA LA GESTION DE RESIDUOS URBANOS SOLIDOS (ATEGRUS), Mugica y Butron 10–2°, E 48007 Bilbao, Tel: 4 445 24 62.

SWEDEN SVENSKA RENHALLNINGSVERKS-FORENINGEN, Ostergaten 30, S 211 22 Malmö, Tel: 040 10 40 45.

SWITZERLAND ISWA SWISS COMMITTEE, c/o O.R.E.D., Junkerngasse 56, 3000 Berne 8, Tel: 031 22 77 85.

UNITED KINGDOM INSTITUTE OF WASTE MANAGEMENT (I.W.M.), 3 Albion Place, Derngate, Northampton NN1 1UD, Tel: 0604 204 2678.

UNITED STATES OF AMERICA AMERICAN WORKS ASSOCIATION—INSTITUTE OF SOLID WASTES (A.P.W.A./ I.S.W.), 1313 East 60th Street, Chicago, Illinois 60637, Tel: 312 667 22 00.

ISWA Reference Addresses

- General Secretariat

Mr. Michel Brès
Association Générale des Hygiénistes
et Techniciens Municipaux (AGHTM)
9, rue de Phalsbourg,
75854 PARIS CEDEX 17 – FRANCE
Tel. (1) 42 27 38 91

- Chicago Subsecretariat

Mr. William S. Forester
Institute for Solid Wastes
American Public Works Association
1313 East 60th Street
CHICAGO, Illinois 60637 – USA
Tel. 312 667 2200
Telex 4949201 APWA

- Scientific Secretary

Prof. Luca Bonomo
Istituto di Ingegneria Sanitaria
Politecnico di Milano
Via Fratelli Gorlini 1
20151 MILANO MI – ITALY
Tel. 2 45 20 354
Telex 333467 POLIMI I

- Waste Management and Research

Editorial Office
Ingeniørhuset,
Vester Farimagsgade 29
DK – 1606 COPENHAGEN V – DENMARK
Tel. 1 156565

Preface

The book is presented to you with the aim of answering a question which often worries not only ISWA members but also wastes specialists throughout the world: How is the waste problem considered in other countries, how is it treated and who does it? We thought that providing this information to ISWA member-countries would be the first step and we have tried to take it.

We have partially succeeded in doing so. This first version remains incomplete and imperfect. Only seventeen countries participated in the work and their chapters are uneven in contents as well as in style. We will try to do better in four years' time. However, the good point is its presence.

I wish to thank all those who contributed to its achievement, beginning with all those who are well-known or anonymous who have collected information and written the chapters. Above all, I want to thank and congratulate the two main authors: first, Mr. Bonomo, Scientific Secretary of ISWA, who had the idea of doing it, defined its structure, examined the chapters and put them into words; then Mr. Higginson who was given the thankless task of reading through all the text to correct omissions in the translations. I hope the success of this book will reward all those who took part in it.

Jean DEFECHE
ISWA President

Contents

1

Presentation

The International Solid Wastes and Public Cleansing Association (ISWA) is a technical-scientific group coordinating competences for managers, researchers, administrators and policy makers working in the field of waste management. Founded in 1970, there are at present twenty-five national members. Among them are nearly all the European countries, including both those based on a market economy and on a planned economy, some important American countries (Argentina, Brazil, Canada, United States of America), Israel, Japan and South Africa.

According to its statute, the objectives of ISWA are to promote adoption of acceptable systems of solid wastes management and of public cleansing through technological development and improvement of practices for environmental protection and conservation of materials and energy resources.

ISWA intends to pursue these objectives by:

- The exchange of knowledge and experiences in the collection, treatment, recycling and disposal of domestic, commercial and industrial wastes, in solid or sludge form, as well as in street cleaning, snow removal and use of related equipment.
- Keeping abreast of, and, where appropriate, promoting scientific research and development in the matter described in the preceding paragraph with coordination between various countries.
- Making available results of research, technical developments, new analytical and working procedures, etc., as well as related practical applications in scientific and technical publications issued if possible in several languages.
- Providing advice on all aspects within the field of activity of ISWA. However, the activities of ISWA do not include giving expert opinion.
- Organizing congresses, seminars, symposia, exhibitions or similar gatherings on an international or regional level; cosponsoring such meetings organized by other associations provided it is in the interest of ISWA.
- Cooperating with industry with a view to develop and improve technologies and knowledge regarding plant and equipment required in the field of activity of ISWA.
- Establishing and maintaining relations with other international or

national organizations whose activities are related to solid wastes management, public cleansing and protection of the environment.

Actually, according to its statute, ISWA—by its central and peripheric structures—works essentially in the following fields:

(a) Scientific organization of conferences and seminars, both by direct involvement of the Scientific and Technical Committee and Working Groups and by various forms of sponsorship. The most important gathering is the ISWA Congress, coupled to an exhibition of materials and equipment, occurring every four years. The ISWA Congress covers the entire organization's professional interest, providing the opportunity for international communications on waste management and research. So far, such Congresses have been organized in Prague (1972), Padova (1976), London (1980) and Philadelphia (1984). The next one will take place in Copenhagen in 1988. Several more conferences are organized on a regional basis or on specific subjects. Four years after the Congress of Philadelphia, we can recall their wide international resonance and specificity of their subjects:

- The two seminars held in Copenhagen on incinerator emissions of heavy metals and particulates (1985) and of trace organics (1987), supplying actual information on the problem of the emissions from municipal solid waste incinerators;
- The symposium in Paris on the upgrading of wastes, costs, recovery and marketing (1986), also dealing with the economic and commercial aspects related to the various techniques for the upgrading of materials and energy recovered from wastes;
- The symposium in Tokyo on hazardous waste (1987); its problems of collection, transportation, treatment and disposal both in industrialized and developing countries;
- The symposium in Cagliari (1987) on process, technology and environmental impact of sanitary landfill.

(b) Publication of printed material, including, besides the conference documents, the journal and the newsletters. The journal (Waste Management and Research), issued quarterly since 1982 by Academic Press, deals with basic research, technical and management aspects. The editors are assisted by an editorial board composed of managers, scientists and policy makers selected according to their professional competence, language and origin. The journal—and most of the conference documents—is written in English although in particular cases, papers in French, German or Spanish may be accepted. The newsletters have the chief task to inform about the association activities. Relevant matters noted are the ISWA meeting calendar, reports from working groups, editorials from ISWA officers, minutes from ISWA meetings, preview of Waste Management and Research contents.

(c) The Scientific and Technical Committee (STC) and working groups activity. STC coordinates all the scientific and technical activities of the association, defining the programmes to be submitted to the governing board. The working groups, on the contrary, are established in dealing with specific subjects, to carry out thorough studies and to acquire new knowledge from the interaction between experienced researchers and managers from several countries. The results appear either in publications as individual reports or in journals or in seminars and conferences. At present, the following topics are being studied:

- Hazardous wastes. The working group and the representatives of 17 countries, have been responsible for the organization of the 1987 symposium in Tokyo and the publication of a report about the status of hazardous waste management in different countries (International Perspectives on Hazardous Waste Management, published by Academic Press, 1987);
- Sanitary landfill. Research in this topic has been more recent and has led to the organization of the first symposium in Cagliari, Italy and also a specialized seminar to be held in Amsterdam in September 1988 as a follow-up of the ISWA Congress in Copenhagen.

The present volume could be included in the above-mentioned publication activity. Its aim is not to deal with technical or scientific aspects which is adequately discussed in the journal and conference documents, but rather to supply global information on the present policies and tendencies and to provide references on the associations, institutes, administrative and technical bodies operating in the waste management field, as well as recent implementations of particular technical interest.

The various chapters, one for each of the member countries of the association, contain the following points:

(a) Introduction on the present national policy in the field of waste disposal and on the evolution trend;
(b) Addresses of the government offices dealing with the problems of municipal and hazardous waste disposal, the technical and professional associations and the main research institutes;
(c) Information on the scientific and technical journals and conferences and exhibitions which take place regularly;
(d) Statistical data on the production and composition of municipal solid wastes and the systems used for their collection and disposal;
(e) Some information on industrial wastes;
(f) Information on the most interesting recent implementations, with addresses;
(g) Reference law framework.

Purposely, the volume does not deal more than marginally with the problem of hazardous wastes, which is being studied by the working group. It has recently published a specific report.

Each chapter has been drawn up by the national member of the association. Fourteen European countries, Canada, Japan and United States of America have supplied their contribution to the work. Some dishomogeneity may exist among the various chapters, partly due to true peculiarities in the situations of the different nations. We hope they will be reduced in the following editions of the volume, since the association intends to update it at regular intervals, including those member countries who are not present now.

Luca Bonomo
ISWA Scientific Secretary

2

Austria*

Surface: 83 855 km^2
Population: 7 565 603 inhabitants (1986)

Municipalities: 2304 (January 1987)

- More than 1 million inhabitants: 1 (Vienna)
- Between 100 000–1 000 000: 4
- Between 10 000–100 000: 61
- Less than 10 000 inhabitants: 2238

1.0 Waste Policy

A change of the Federal Constitution is expected to give the federation authority for the entire field of waste.

The waste management strategies are to be arranged by the federation, the counties and the municipalities depending on their regional aspects.

Initial scientific–technical principles and aims

Waste management strategies have to be organized in such a way that only two kinds of substances, namely reusable and inert ones, are produced.

Methods of waste treatment are devised in such a way that substances which endanger the environment are produced in the most concentrated form and substances which do not endanger the environment are produced in the most inert form.

The final treatment of waste must take place in Austria.

Important economical principles

Waste materials have to be recycled to reduce pollution especially from the final treatment of wastes and the corresponding production of new materials. Recycling systems should be supported by fees paid compulsorily for the final treatment if an operational profitability of these recycling systems does not exist.

*By dipl.ing.P.Lechner, mag.P.Mostbauer, dipl.ing.R.Pawlick – Technische Universität Wien; doz.dr.G.Vogel,dr.Mayr, mag.Krenn – Wirtschaftsuniversität Wien.

2.0 Addresses

2.1 National member of ISWA

ISWA—Austrian Section
c/o Österreichischer Städtebund
Rathaus
A-1082 Wien

- National representative and executive committee member: Senatsrat Dipl.Ing. Gilnreiner, Magistratsabteilung 48, Einsiedlergasse 2, A-1050 Wien.
- Scientific and Technical Committee member: Prof. Dr. Walter Kemmerling, Technische Universität Wien, Institut für Wassergüte. Karlsplatz 13, A-1040 Wien.

2.2 Government offices

- Bundesministerium für Umwelt, Jugend und Familie (Ministry of the Environment, Youth and Family).
  A-1030 Vienna, Radetzkystr. 2, Tel. 0222 75 56 86 0

Competence: general environmental protection and initiating coordination; nature conservation, maintenance of unpolluted air, noise control, waste disposal, radiation protection, consumer protection, implementation of the Hazardous Waste Act, as far as Federal competence is concerned.

- Department I, I/1 . . I/5: Environmental Affairs
- Umweltbundesamt (Federal Environment Agency)
  A-1010 Vienna, Biberstraße 11, Tel. 0222 512 35 21 0
- Öko-Fondskommission (Ecology Funds Commission)
  A-1030 Vienna, Reisnerstraße 4, Tel. 0222 72 51 07 0
- Österreichisches Bundesinstitut für Gesundheitswesen (Federal Institute of Public Health)
  A-1010 Stubenring 6, Tel. 0222 515 61 0

- Bundesministerium für wirtschaftliche Angelegenheiten (Ministry of Economic Affairs)
  A-1010 Vienna, Stubenring 1–3, Tel. 0222 75 00 0

Competence: maintenance of unpolluted air, noise control, waste disposal, new (clean) technologies.

- Department I/4a: General Economic Policy: Coordination of Environmental Protection Affairs.
- Department III/11: General and legislative regulations of factory law with regard to environmental protection, cooperation with the Ecology Funds Commission.
- Department IV/4: Industrial Policy with regard to environmental protection.
- Department VII/2: Research and Development in the field of secondary raw material utilization (recycling).
- Department X/2: Waste avoidance, hazardous waste and implementation of the Waste Oil Act (1986).

- Bundesministerium für öffentliche Wirtschaft und Verkehr (Ministry of Public Economy and Transport)
  A-1010 Vienna, Radetzkystraße 2, Tel. 0222 75 56 86 0
  - ERP-Fonds, A-1010 Vienna Renngasse 5, Tel. 534 64 0
  - Department V/13: Cooperation in the field of environmental protection technologies.

- Bundesministerium für Wissenschaft und Forschung (Ministry of Science and Research)
  A-1010 Vienna, Freyung 1, Tel. 0222 531 20 0
  - Department II/3: Raw material research.
  - Department II/4: Recycling research.
  - Department II/5(25): Environmental research.
  - Department II5(25)a: Energy-related environmental technology, Government Program: "Environmental Technology".
  - Department III/VII/4: Coordination of hazardous waste affairs.
  - Department III/VII/5: Waste and waste material utilization.

2.3 Organizations

- Bundeskammer der gewerblichen Wirtschaft (Federal Chamber of Trade and Industry)
  A-1040 Vienna, Wiedner Hauptstraße 63, Tel. 0222 501 05 0

  Arbeitgemeinschaft Umweltschutz (Joint Committee of the Federal Chamber of Trade and Industry, the Industrialists' Association, and of the Presidents' Committee of the Austrian Chamber of Agriculture).

  Information and coordination of associates; technical, economic and juridical first-hand advice for members as to factory approvals, factory planning, process technologies, noise, waste and energy problems, etc.

- Bundesabfallbörse bei der Kammer der gewerblichen Wirtschaft für Oberösterreich (Federal Waste Exchange of the Chamber of Trade and Industry of Upper Austria)
  A-4010 Linz, Hessenplatz 3, Tel. 0732 26 75 35

Information of secondary materials supply and offer at home.

- Österreichisches Normungsinstitut (Austrian Institute of Standards)
  A-1020 Vienna, Heinestraße 38, Tel. 0222 26 75 35 0
  Elaboration of Austrian Standards in the fields of air-measuring technology, maintenance of unpolluted air, waste control and disposal, water, sewage water, radiation protection and nuclear technology, information and sale of Austrian Standards.

- Abfallsammel- und Verwertungsagentur (ASVA) (Waste Collection and Utilization Agency),
  A-1063 Vienna, Mariahilferstraße 77–99, Tel. 0222 96 36 11 0
- Österreichische Gesellschaft für Natur- und Umweltschutz (Austrian Society for Environmental Protection) (ÖGNU), Hegelgasse 21, A-1010 Wien
- Ökologie-Institut, Neubaugasse 64–66, A-1070 Wien
- Niederösterreichische Umweltschutzanstalt, Südstadtzentrum 4/4, A-2344 Maria Enzersdorf

2.4 Research institutes

- Technische Universität Wien (University of Engineering and Technology)
  Institute of Water Quality, o. Univ. Prof. Dr. Walter Kemmerling and Dipl. Ing. Peter Lechner
  Karlsplatz 13
  A-1040 Vienna
- Technische Universität Wien (University of Engineering and Technology)
  Institute for Process Engineering and Technology of Fuels, a.o. Univ. Prof. Dr. Albert Hackl
  Getreidemarkt 9
  A-1060 Vienna
- Wirtschaftsuniversität Wien (University of Economics and Business Administration)
  Institute of Technology and Commodity Science
  Assistant Professor Dr. Gerhard Vogel

Augasse 2–6
A-1090 Vienna

- Technische Universität Graz (University of Engineering and Technology)
  Technical Environmental Protection (Postgraduate Studies)
  o. Univ. Prof. Dipl.-Ing. Dr. Rolf Jürgen Marr
  Rechbauerstraße 12
  A-8010 Graz

3.0 Spreading of Information

3.1 Scientific–technical journals and series

- Der Aufbau, magazine for planning, construction, housing and environmental protection.
- Umweltschutz, the Austrian magazine for ecology, economy and environmental technology.
  Editor: Bohmann druck und Verlag, Vienna
- Publication series "Abfallwirtschaft". Editor: Technische Universität Wien, Institut für Wassergüte und Landschaftswasserbau, Abteilung Abfallwirtschaft
- Publication series "Wiener Mitteilungen". Editor: Technische Universität Wien, Karlsplatz 13, A-1040 Wien, Universität für Bodenkultur Wien
- Publication series
  Editor: Institut für Technologie und Wasserwirtschaftslehre, Wirtschaftuniversität Wien

3.2 Exhibitions, conferences and seminars

- Abfallwirtschaftsseminar TU Wien, Seminars on Waste management. Organized by Institut für Wassergüte und Landschaftswasserbau, Abteilung Abfallwirtschaft, Technical University of Vienna
- Grazer Seminar Regionale Abfallwirtschaft (Seminar on Waste Management) (September 17–18, 1987, Graz/Styria)
  - Integrated Waste Management Plans
  - New Technologies of Waste Processing in the Framework of Waste Management Plans
  - Waste Management and Environment
  - Rentability of Integrated Waste Management Plans

- Ökologia
  International Exhibition and Congress on Environmental Affairs (November 7–15, 1987, Vienna) organized by the Wiener Akademie für Zukunftsfragen (Austrian Academy for Questions of the Future, A-1010 Vienna, Wollzeile 12) in cooperation with the Committee "Save the Woods" and the Ministries for the Environment and Science and Research, the Ecology Funds, the Federal Environment Agency and the International Society for Environmental Protection.
- UTEC–1. Kongress-Messe für Umwelttechnik (First Exhibition and Congress on Environmental Technology (December 3–5, 1987, Linz/ Upper Austria)
  Special sessions on
  - Waste management planning
  - Heat-treatment processes in waste utilization
  - Landfill technology and composting
  - Practical cleansing measures
  - Measuring techniques
  - Maintenance of unpolluted air
  - Water and sewage treatment
  - Medicine and environment
  - Noise control
  - Construction biology
  - Optimum energy utilization
  - Management and financing
  - Environmental legislation

3.3 University teaching

- Technische Universität Wien (University of Engineering and Technology)
  Institute of Water Quality, Department of Waste Management. Head: Prof. Dr. Walter Kemmerling and Dipl.Ing. Peter Lechner
  Teaching, carrying out research studies and giving expert advice in the whole field of waste management.
  Karlsplatz 13
  A-1040 Vienna
- Technische Universität Wien (University of Engineering and Technology)
  Institute for Process Engineering and Technology of Fuels, a.o. Univ. Prof. Dr. Albert Hackl
  Technical Environmental Protection (Postgraduate Studies)

Getreidemarkt 9
A-1060 Vienna

- Wirtschaftsuniversität Wien (University of Economics and Business Administration)
  Institute of Technology and Commodity Science
  Assistant Professor Dr. Gerhard Vogel
  Information and expert advice on management plans, environmental protection problems, especially in the field of solid waste management and the carrying out of scientific research studies (waste analyses etc.)
  Augasse 2–6
  A-1090 Vienna
- Technische Universität Graz (University of Engineering and Technology)
  Technical Environmental Protection (Postgraduate Studies)
  o. Univ. Prof. Dipl.-Ing. Dr. Rolf Jürgen Marr
  Rechbauerstraße 12
  A-8010 Graz
- Universität für Bodenkultur (University of Soil Sciences)
  a.o. Univ. Prof. Dipl. Ing. Dr. Werner Lengyel
  Gregor Mendel Straße 33
  A-1180 Vienna
- Universität Innsbruck (University of Innsbruck/Tyrol)
  o. Univ. Prof. Dipl. Ing. Dr. Ingerle
  Innrain 52
  A-6020 Innsbruck
- Off-university courses:
  WIFI (Wirtschaftsföderungsinstitut der Kammer der gewerblichen Wirtschaft) (Institute for Economic Development with the Chamber of Commerce and Trade)
  Währinger Gürtel 97
  A-1181 Vienna
  12-day seminar on "Waste treatment in commercial enterprises".

4.0 Production and Composition of Wastes

4.1 Categorization

According to the Austrian Standard ÖNORM S 2001, waste management is defined as the total of all measures concerning waste and their purposeful coordination with regard to their influence on the environment and their rentability.

According to ÖNORM S 2000 waste is categorized as follows:

Waste

Solid Waste:
- Household waste
- Industrial waste
- Bulky waste
- Street sweepings

Special and hazardous waste:
- Special waste
- Hazardous waste
- Extremely hazardous waste

4.2 Waste production in Austria

According to an investigation of the Austrian Federal Institute of Public Health (ÖBIG) in 1984 the annual waste production in Austria amounted to 2 158 000 t and the production of hazardous waste to 200 000 t (Table 1).

Table 1 Sweepings, household, commercial and industrial waste in Austria

Waste Category	Weight (1984) ($\times 10^3$ t)
Sweepings, household, commercial and industrial waste amounts collected	1630
Bulky waste amount collected	98
Direct deliveries to waste treatment plants of household, commercial and industrial wastes	328
Separate collection of secondary materials	102
Total	2158

Table 2 Household solid waste amounts in Austria

Material	Weight (%)	Weight ($\times 10^3$ t)	Specific amount (kg/p/year)
Glass	11.1	181	24
Wood, leather, rubber	1.8	29	4
Plastics	6.0	98	13
Metals	8.4	137	18
Minerals	15.1	246	33
Paper and paperboard	27.2	443	59
Textiles	7.9	129	17
Organic materials	22.5	367	49
Total	100.0	1630	217

Table 3 Austrian waste composition 1973–1986 (in percentage weight)

Material	1	2	3	4	5	6	7	8	9	10	11
Glass	11.1	9.2	8.3	8.4	9.5	10.3	5.1	6.4	11.0	9.7	8.6
Wood, leather, rubber	1.8	2.2	2.2	0.3	0.8	1.1	3.4	3.8	2.1	2.0	0.4
Plastics	6.0	6.1	8.8	3.8	7.0	9.2	11.1	10.5	7.2	6.8	8.7
Metals	8.4	8.1	3.4	2.6	3.3	4.5	5.1	6.9	3.7	3.8	3.9
Minerals	15.1	9.9	5.7	6.4	19.4	15.5	13.5	12.0	11.4	15.2	10.8
Paper and board	27.2	38.3	32.75	61.9	30.5	23.7	36.2	37.0	34.1	33.2	32.9
Textiles	7.9	7.6	5.5	1.0	2.8	5.2	4.0	3.6	2.5	3.7	2.2
Organic materials	22.5	18.6	31.2	15.6	26.7	30.5	21.6	19.8	24.2	22.3	32.5
Compounds	—	—	2.2	—	—	—	—	—	2.9	2.6	—
SQHW*	—	—	—	—	—	—	—	—	0.9	0.7	—
Total	100.0	100.0	100.0	100.0	100.0	100.0	100.0	100.0	100.0	100.0	100.0

* Small Quantity Hazardous Wastes.

Where 1: Austria 1973
2: Vienna 1973
3: Vienna, Per Albin Hansson-Siedlung 1977/78
4: Vienna, offices 1980
5: Freistadt, OÖ, 1981
6: Altenberg, OÖ, 1981
7: Vienna 1982
8: Vienna 1982/83
9: Vienna, 19th district, 1985/86
10: Vienna, 20th district, 1985/86
11: Graz, 1985/86

Table 4 Austrian waste composition 1973–1986 (in percentage volume)

Material	1	2	3	4	5	6	7	8	9	10	11
Glass	5.2	3.8	2.8	1.4	5.7	7.1	—	—	3.2	2.8	—
Wood, leather, rubber	2.3	2.9	2.1	0.2	2.3	2.7	—	—	2.8	2.5	—
Plastics	18.8	14.5	19.7	5.2	15.6	21.1	—	—	21.6	21.5	—
Metals	9.5	8.4	3.5	2.6	5.4	7.2	—	—	3.6	3.3	—
Minerals	3.9	2.2	2.7	4.7	8.3	10.9	—	—	3.7	3.9	—
Paper and board	40.2	52.2	42.2	76.8	43.5	28.9	—	—	42.7	45.4	—
Textiles	8.4	7.2	6.9	0.7	3.7	9.0	—	—	3.3	4.3	—
Organic materials	11.7	8.5	15.8	8.4	15.5	13.1	—	—	11.6	9.3	—
Compounds	—	—	4.3	—	—	—	—	—	7.3	6.9	—
SQHW*	—	—	—	—	—	—	—	—	0.3	0.2	—
Total	100.0	100.0	100.0	100.0	100.0	100.0	100.0	100.0	100.0	100.0	100.0

* Small Quantity Hazardous Wastes.

Where 1: Austria 1973
2: Vienna 1973
3: Vienna, Per Albin Hansson-Siedlung 1977/78
4: Vienna, offices 1980
5: Freistadt, OÖ, 1981
6: Altenberg, OÖ, 1981
7: Vienna 1982
8: Vienna 1982/83
9: Vienna, 19th district, 1985/86
10: Vienna, 20th district, 1985/86
11: Graz, 1985/86

4.3 Waste composition

The composition of Austrian household waste was first investigated in 1973 in the framework of a small representative study. Other investigations followed later, in the framework of research studies for a separate collection of secondary materials (Tables 2, 3 and 4).

5.0 Waste Collection in Austria

The standard of waste collection in Austria is very high, since nearly all inhabitants are reached by a collection organization, as shown in Table 5. Weekly collection depends on the number of persons. However, in a number of communities regardless of the number of persons, collection every other week prevails.

Table 5 Household waste collection in the 2301 Austrian communities

Collector	Inhabitants reached (%)	Communities reached (%)
Community or Community district	58.6	28.4
Private collection company	40.1	67.5
Households	1.3	4.1
Total	100.0	100.0

Table 6 Container sizes used in household waste collection

Container size (l)	Share in the available container volume	
	Austria (Vol. %)	Vienna (Vol. %)
Bucket/bag, >60	4.0	—
Container/bag, 61–120	31.9	12.5
Container, 121–240	10.3	17.9
Large container, >240	53.8	70.6
Total	100.0	100.0

5.1 Some different waste collection and treatment systems in Austria

Vienna

Vienna (1 555 000 inhabitants) has two large incineration plants with wet scrubbing systems. The energy is used all through the year for district heating.

The goals of the new waste management plan are:

1.0 Waste minimization by quality and quantity;
2.0 Recycling and resource recovery;
2.1 Recycling of secondary raw materials;
2.2 Composting of the separately collected organic fraction of household waste;
2.3 Energy recovery of wastes;
3.0 Environmental sound disposal of the residues from all treatment plants;
4.0 Minimization of the costs and the influence on the environment of all systems elements.

The new intensive separate collection system of waste paper, waste glass and some other secondary raw materials reduces the amount of wastes by 22% in weight and an additional 16% in the composting activities in some areas.

Graz

The City of Graz principally pursues a similar waste management plan to Vienna. Due to political, meteorologic (inversion weather) and energy policy reasons (insufficient district heating), incineration of mixed waste is not desirable. Therefore, secondary and hazardous materials are collected separately and the residual waste is brought to a waste treatment plant (mechanical separation set up by the VOEST). This divides the waste into three fractions (apart from ferrous scrap separation): RDF-fraction (38%), compost raw fraction (40%) and a hard material fraction (22%). At present, the three fractions are landfilled at three different landfills, but outlets are sought. The VOEST has presented to the City, an offer to construct a high temperature gasification plant. The city of Linz has a prototype of such a unit.

Wels

Apart from Vienna, Wels (Upper Austria, 92 000 inhabitants) is the only city in Austria that has an incineration plant for the heat treatment of waste. The waste from about 55 100 people is incinerated there. Unlike the Viennese plants, the only purpose of the incineration in Wels is to reduce the waste amount to be landfilled and not to be utilized for district heating.

Flue gas cleaning is effected by electrostatic precipitators only. Scrap is not previously separated.

Community District Kirchdorf a.d. Krems

The central element of waste treatment of the community district Kirchdorf a.d. Krems, Upper Austria, is a separation unit with magnetic separation of ferrous metals and a sieve-classification of the fractions RDF, compost raw materials and landfilling material. Twenty-two communities with about 50 000 inhabitants (1240 km^2) of the community district and one neighbouring community with 1400 inhabitants are supplied. At present the separated RDF is not utilized.

District Neunkirchen

Since January 1986, the district Neunkirchen, Lower Austria, with 86 000 inhabitants (1146 km^2) is cleansed according to the system "Grüne Tonne", with one container for a dry solid waste mixture and one for the remaining wet waste.

The solid waste mixture is separated in fractions composed of paper, glass, ferrous metals and NE-metals, textiles, plastics and styrofoam; compost raw materials are separated from the wet waste container. Parts of both material streams are processed to RDF.

At present, there is no possibility of incinerating the obtained RDF—the fluid bed incineration plant of the paper factory W. Hamburger AG, Pitten, that holds a 50% partnership, has only a license to incinerate residues from waste paper processing—so that the RDF has to be landfilled. There are similar difficulties in finding outlets for different secondary materials after the separation.

The ratio of dry waste to wet waste is 42 : 58, with a specific waste amount of 215 kg/p/a. The RDF share amounts to about 9.4%.

Salzburg

The composting plant in Siggerwiesen, in operation since 1978, is planning heat utilization for remaining materials. It is not yet clear whether this project will be carried out in the framework of the energy policy plans of the City of Salzburg (extension of district heating).

5.2 Acceptance of RDF and secondary materials by the Austrian industry

Many of the new investments in the Austrian industry for boiler plants include fluid bed incineration. It is known that this technique can utilize RDF and secondary materials to a certain extent. Forthcoming legislation,

Table 7 Comparison of material streams of selected Austrian waste management plants with heat treatment of waste (in weight percentage, Kirchdorf in volume percentage)

Location	Waste materials with scrap	Raw compost	Waste incineration	Landfilling	Total
Vienna					
state 1986	12%	—	36%	56%[1]	100%
planned state	22%[2]	5%[3]	41%	32%[1]	100%
Wels	8%[4]	—	49%	43%	100%
Grüne Tonne					
Neunkirchen	21%[5]	34%	9%[6]	36%	100%
Graz	9%[7]	36%	35%	20%	100%
Kirchdorf[8]	8%	20%	47%	25%	100%

Anm.: [1]Including residues from waste incineration.
[2]From separate collection and separating units.
[3]From separate collection with biocontainer.
[4]Conventional separate collection.
[5]Secondary materials partly landfilled at present.
[6]RDF landfilled at present.
[7]4%secondary materials from separate collection, 5% scrap from waste treatment plants.
[8]Data in volume percent.

however, will establish the same emission limit values for RDF incineration as for waste incineration, i.e. RDF utilization will only be possible with an adequate gas scrubbing unit. On the other hand, incineration of RDF and secondary materials in either fluid bed incinerators or concrete rotary kilns is an adequate waste management measure in a relatively low populated area as Austria.

Table 8 Waste treatment in Austria

Type	Number	Cleansing area: Persons ($\times 10^3$)	Cleansing area: (%)	Waste amount treated: ($\times 10^3$)	Waste amount treated: (%)
Composting and rotting plants	17	2114	27.9	492.5	24.0
Incineration plants	3	986	13.1	335.0	16.3
Landfills (household and bulky waste)	517	4455	59.0	1228.0	59.8
Total	537	7555	100.0	2055.5	100.0

Table 9 Cost and fees of waste treatment in Austria

Container category	Fee/emptying: Most frequent values (S)	Fee/emptying: Range (S)	Density (t/m^3)	Costs (S/t)	Costs (S/m^3)
1201	25–30	5–75	0.154	1.488	230
2401	50–60	10–80	0.135	1.081	229
11001	160–200	100–320	0.109	1.054	164

6.0 Treatment and Disposal

The waste treatment plants presently operating are shown in Table 8.

7.0 Treatment and Disposal of Municipal Solid Waste

7.1 Material recovery

In Austria, there are two integrated collection systems for secondary raw materials, namely for waste paper and waste glass. In comparison to 1986, when 77 040 t of waste glass were collected in Austria, the amount collected in 1987 was 84 600 t, corresponding to an increase of 9.8%. This kind of material recovery reduced the total waste amount in 1987 as follows:

- 254 000 m^3 in collection containers
- 110 000 m^3 in collection vehicles
- 58 000 m^3 at landfills

Waste paper collection in Austria was carried out in about 90 000 collection containers and amounted to 85 000 t of waste paper, in 1987 according to the Austria Recycling 102 125 t of waste paper (without Vorarlberg). The waste paper amount collected in 1986 corresponds to about 23% of the total waste paper amount in Austria. The rest is produced by the Austrian commerce. Apart from waste paper and glass collection, about 2000 t of old textiles and an unknown amount of old tyres, metals, cars, car batteries and plastics are utilized.

7.2 Composting

A great deal of interest in composting activities was shown during the 1970s and the beginning of the 1980s—17 composting plants were established and

Table 10 Austrian composting plants

Name/location	County	Starting	Served inhabitants ($\times 10^3$)
Attnang/Redlham	Oberösterreich	1975	173
Lustenau	Vorarlberg	1975	250
Feldkirchen/Donau	Oberösterreich	1976	143
Katsdorf	Oberösterreich	1976	110
Traiskirchen	Niederösterreich	1976	96
Aich-Assach	Steiermark	1978	30
Allerheiligen	Steiermark	1978	112
Oberpullendorf	Burgenland	1978	272
Pöchlarn	Niederösterreich	1978	72
Siggerwiesen	Salzburg	1978	350
Zell am See	Salzburg	1978	100
Taufkirchen/Trattnach	Oberösterreich	1979	56
Frojach-Katsch	Steiermark	1981	85
Liezen	Steiermark	1981	56
Roppen	Tirol	1985	62
Inzersdorf	Oberösterreich	1986	50
Neunkirchen	Niederösterreich	1986	97

are still working. 2.1 million inhabitants are served by these plants, as shown in Table 10.

In the 1970s nearly 50% of the MSW was converted into compost. The mixed household waste was crushed and composted without any preparation. During the composting period, contact with heavy metals from the waste products occurs. The longer this contact period takes place, the more metals will dissolve and degrade.

Table 11 ÖNORM-standards for compost–organics and nutrients

		% dry substance
Organics		18–30
Nitrogen	(N)	0.5–1.5
Phosphorus	(P_2O_5)	0.4–0.8
Potassium	(K_2O)	0.3–1.0
Calcium	(CaO)	2.0–12.0
Magnesium	(MgO)	0.5–3.0
Iron	(Fe)	1.0–6.0
Manganese	(Mn)	0.05–0.12

Table 12 Heavy metal content in composts and soils in Austria (ppm dry substance)

Metal		ÖNORM S 2022 1984	Proposal for ÖNORM revision 1987	Composted from mixed household waste (32 composts 1984–1986)	Compost from source separated kitchen waste 1987	Limits for soil for agricultural use in Austria 1987	Frequency in Austrian soils 1987
Chromium	(Cr)	≤300	≤150	98	12	≤100	2–50
Nickel	(Ni)	≤200	≤100	80	22	≤60	2–50
Copper	(Cu)	≤1000	≤500	333	57	≤100	1–20
Zinc	(Zn)	≤1500	≤750	1450	135	≤300	3–50
Cadmium	(Cd)	≤6	≤3	5.0	<1	≤2	0.01–1
Mercury	(Hg)	≤4	≤2	n.d.	n.d.	≤2	0.01–1
Lead	(Pb)	≤900	≤450	728	40	≤100	0.01–20

n.d.: not determined.

To protect the environment against this bad compost, a quality standard was set by an ÖNORM (ÖNORM S 2022—Quality requirements for waste compost/1984). This ÖNORM sets standards for organic substances like nutrients, physical properties, heavy metals, and inert substances such as glass, metals and plastics and also for the effect on plant germination and growth.

In 1985, the analytical methods and quality control for waste compost were also fixed by ÖNORM (ÖNORM S 2023). In the case of agricultural soils, the level of heavy metals is limited by law since 1987.

In the same year a guide for application of waste compost was established (ÖNORM S 2024).

Concerning the rate at which the heavy metal content of soil can increase the dosage of compost with given quality is therefore limited.

Table 13 Maximum dosage of heavy metals on agricultural land

Cr	1250 g/ha year
Ni	250 g/ha year
Cu	1250 g/ha year
Zn	5000 g/ha year
Cd	25 g/ha year
Hg	25 g/ha year
Pb	1250 g/ha year

The consequences for running composting plants are:

1. To diminish the heavy metal content by source separation of hazardous waste from households and by carrying out processing methods, which results in an organic fraction with a low heavy metal content.
2. The compost has to fit quality standards regarding the effect on germination and growth of plants. Therefore it is necessary to reflect on the subsequent biogradation systems.

At the moment all seventeen composting plants are still running. They are processing a total of about 360 000 t MSW per year, the compost output is 70 000 t per year. Compared to the 1970s, the percentage of compost production is dramatically reduced to 20% of the MSW-Input, but the heavy metal content now fits the ÖNORM in most cases.

As it is impossible to overcome all problems of undesirable materials and heavy metals from mixed waste, a combination of source separation of recyclable and organic waste with subsequent composting is thought to be the actual goal. In this case, a high quality of waste is diverted from landfilling, but all necessary steps must be taken to make sure that no heavy metals enter the organic components.

Source separation of only recyclables (paper, glass and metals) was started in the early 1970s, but now in many small villages (e.g. Schwechat, Klosterneuburg) and also in big towns (e.g. Vienna) an additional source separation of organics with subsequent composting is projected.

7.3 Incineration

About 20% of Austria's municipal refuse and a considerable portion of Austria's hazardous waste can be combusted by four incineration facilities.

Table 14 Austrian incineration facilities–capacities and other characteristics

Name/location	Starting	Capacity (t/year × 10^3)	Waste/type	Position
Flötzersteig	1963	180	Municipal refuse	Urban
Spittelau	1970	210	Municipal refuse	Urban
Wels	1973	32	Municipal refuse	Peripheral
EBS-Simmering	1974	≃100	Hazardous waste	Peripheral
EBS-Simmering		70	Waste water treatment sludge	

Table 15 Limits for flue gas emissions of waste in Austrian incineration facilities

Parameter (mg/m^3)	National emission standard	Regional standard EBS-Simmering	Observed values January 1988 EBS-Simmering rotary kiln ovens
Dust	25	10	0–2
Hydrochloric acid	30	15	<3
Hydrofluoric acid	0.7	0.1	<0.1
Sulfur dioxide	100	100	<20
Carbon monoxide	100	—	—
Pb + Zn + Cr-compounds	4	Pb 0.5, Zn 0.8, Cr 0.2	n.d.
As + Co + Ni-compounds	1	As 0.2, Ni 0.2	n.d.
Cadmium	0.1	0.05	n.d.
Mercury	0.1	0.05	n.d.
Equivalents of 2.3.7.8.-TCDD	0.0001	—	n.d.

n.d.: not determined.

About 120 000 t of slags and 40 000 t of fly ash remaining from the incineration of wastes have to be disposed of in landfills. Dust emissions have been controlled by electrofilters for several years. In the past 3 years, all incinerators except the small one at Wels have been equipped with wet flue gas scrubbers of a modified LAB EDV-8000 system. All scrubbers have been installed by the Austrian firm Simmering-Graz-Pauker-AG and consist of 2 spray chambers followed by two successive venturi scrubbers. In addition to national air quality standards, there are very stringent technical requirements for the emission control equipment for those incineration units which are located near densely-populated areas.

In May 1987 there was an accident at the "Spittelau" facility in Vienna. The reason for this accident could not be clearly determined, and the Spittelau facility is still out of operation today. Therefore a lot of municipal refuse has to be disposed of by landfill at Vienna. The scrubber system will be rebuilt soon and recent technological developments will be taken into account.

A new concept for the treatment of scrubber waste and waste waters has been developed by the Simmering-Graz-Pauker-AG that will reduce the annual amounts of the hazardous scrubber filter cake to a minimum. A $DeNO_x$-catalyst to reduce NO_x-emissions at the Spittelau facility will also be installed. If this flue gas installation works well, the other facilities could also be equipped with $DeNO_x$-catalysts.

7.4 Sanitary landfilling

Landfilling is a substantial factor of waste management. Even if waste is largely avoided or recycled, some substances which still remain will have to be disposed properly.

In the past, the local conditions of landfills and their hydrogeological aspects were not thoroughly considered. Many landfills which were situated in vacant areas in the past are now in the suburbs or even in the middle of towns as a direct consequence of the spreading of settlements. As a consequence, the population has been increasingly opposed to the emissions of indiscriminate dumpings. The problematic nature of these dumpings leads to the total opposition against sanitary landfills particularly by direct residents. In Austria, the responsibility for the collection and disposal of household waste has rested with the municipal authorities.

In 1984, a total number of 517 (Table 16) municipal landfills were in use. Only in a few cases, a landfill served a whole region or a big town. In many cases, single municipalities had their own small landfill. The number of landfills is now decreasing.

The present strategy ruling landfill practice is to protect the environment

Table 16 Type and number of landfills in Austria (1984)

According to the types of disposed wastes	Number of landfills	Percentage of landfills (%)
Domestic waste	78	15.1
Bulky waste	6	1.2
Domestic & bulky waste	43	8.3
Domestic & bulky waste + other waste materials	290	56.1
Other waste materials except domestic & bulky waste	72	13.9
Without specification of the kind of wastes	28	5.4
Total	517	100.0

from emissions of the disposed waste and this will continue. As we know that we cannot achieve infinite insulation of the waste, we must accept that the waste will interact in the future with the environment. Therefore the alternative measure is to stabilize the landfill by a certain technique in a limited time.

At the present time, guidelines for sanitary landfills are published by the "Bundesministerium für Land- und Forstwirtschaft" and the "Bundesministerium für Umwelt, Jugend und Familie".

The main points of these guidlines are:

- Site lining is required in all cases.
- The proposed lining system must be constructed of two different materials that have appropriate chemical properties and sufficient strength and thickness to prevent failure. The system requires a primary flexible membrane liner of at least 2 mm of synthetic material, chemically resistant to waste and leachate. This primary flexible membrane liner must be in direct contact with a secondary compacted soil of low permeability ($k_f < 10^{-9}$ m s^{-1}). The low permeability soil component must be at least 60 cm of compacted material with saturated hydraulic conductivity of 10^{-9} m s^{-1} or less. This soil liner should be made of three layers of 20 cm.
- The primary membrane liner should be protected by the drainage layer immediately. As an additional protective measure, a fleece can be placed above the membrane liner.
- The lining system slopes towards the leachate collection pipes at a minimum of 3% gradient.
- The drainage system should be able to remove leachate from any point in the facility. The drainage system consists of a gravel layer

and pipes. If the flow through the pipes is restricted, the drainage requirements should be met by the gravel layer alone.

- The pipe size should be 200 mm in diameter, the minimum pipe slope shall be 2% gradient.
- Access should be provided to all parts of the system. This makes the placing of pits necessary so that inspection and maintenance is possible. The distance between access points should be twice the maximum operating distance of the jet cleaning equipment.
- The drainage pipes should be periodically inspected directly using televisual methods.
- The objective must be to stabilize the landfill in a limited period.
- Leachate treatment: treatment by biological and/or by physico-chemical processes and recirculation through the landfill on site should be preferred.
- Removing gas systems should be undertaken during landfill operation.
- An active gas removal system will be necessary.
- Before beginning the landfill operation, a monitoring network must be installed. The wells should be located in such a way that samples can be obtained which show the quality of upgradient groundwater (background wells) and downgradient groundwater adjacent to the landfill at various depths.
- Source separation of recyclables and organics with subsequent composting and/or methane production combines a high degree of waste diversion from landfilling with fewer environmental problems.

8.0 Treatment and Disposal of Hazardous Wastes

8.1 National and regional control

A national regulatory control program was established in 1983 by the Austrian government. Since then, supervision of some special waste streams (oil, radioactive and infectious wastes, iron scrap) was governed by specific acts and decrees.

The Austrian "Sonderabfallgesetz" (Special Waste Act) from 1983 demands that:

- Waste transporters and facilities have permits from regional government. This also applies to the import and export of wastes.
- Amount and type of hazardous wastes which are generated, transported or treated have to be registered by transporters, facilities and regional authorities.

In 1984, a list of 148 specific hazardous wastes was published. Addition-

Additionally, sale and production of new chemicals is controlled by an Act of the national government (Chemicals Act, 1987).

8.2 Treatment and recovery

Facilities that avoid waste water and solid wastes are subsidized by a fund called "Ökofonds". Originally two funds existed: one of them supporting projects regarding mainly waste water treatment and the other one regarding all remaining environmental projects.

There are many small facilities that recycle or recover substances which have been emitted or disposed previously. However, the amount of wastes originated by air pollution prevention and waste water treatment is still increasing. It can be said that most of these wastes are not hazardous but of great importance because of their large annual amounts. Hazardous wastes which cannot be treated by the facilities themselves are commonly delivered to the hazardous waste treatment plant "EBS". There are only a few small treatment plants that neutralize and detoxify inorganic hazardous wastes and one incineration unit for organic wastes.

In 1987, about 60 000 t of organic hazardous wastes were burnt by the "EBS" treatment facility near Vienna. Recently both rotary kiln ovens of this facility have been equipped with wet scrubbers of the Simmering-Graz-Pauker type. First measurements of HCl, SO_x and dust have shown that all the stringent air quality standards are fulfilled.

8.3 Landfilling

In Austria there is an urgent need for industrial waste landfill sites. On top of that, there is no agreement about conceptional and technical items. At least 5 different concepts have been elaborated but have the following in common:

- Wastes should be avoided or reduced if possible.
- Surface impoundments such as pits, ponds and lagoons as well as landfarming are considered to be insufficient disposal methods.
- Only treated or harmless wastes can be disposed of by landfill.
- No liquids will be accepted.
- The site must be geologically qualified.
- Technical standards for lining system, leachate collection and treatment of leachate have to be fulfilled.

Until now, none of these concepts have been realized. Some attempts of regional authorities to find such landfill sites failed because of the uncompromising opposition by residents. Agreement and national collaboration will be necessary to avoid ecological shortcomings.

9.0 Legislation

Austria is a Federal State composed of nine provinces. The Federal Constitutional Law regulates through a general clause, e.g. for household waste, or rather according to the competence for trade and industrial affairs the following competences in waste management.

Table 17 Legislative competences in waste management in Austria

Waste category	Legislation	Execution
Solid waste	Provinces (9 Provincial Laws)	Communities
Hazardous waste	Federal Government (1 Federal Law)	Provinces

9.1 Standardizing waste management legislation

At present a constitutional amendment concerning the law on waste and waste management is being prepared. Until now the Austrian Provinces have been charged with legislation, administration and management of waste. Thus, a number of different regional waste management acts are in effect.

The creation of a new competence enables the Federal state of Austria to standardize these different regulations. The amendment of the constitutional law will be passed this year.

9.2 Legislation on hazardous waste

The Hazardous Waste Act, BGBI. 186/1983 includes regulations for registration and disposal of hazardous waste as well as executive regulation of the Federal Ministry of Health and Environmental Protection with the exception of all cases for which acts have already been passed.

- The Waste Oil Act, BGBI. 138/1979, with regulations for the treatment of waste oil suited for energy recovery, oil refinement or recycling.
- The Austrian Standard ÖNORM S 2100 (Special Waste Index) lists the substances which qualify as special wastes in Austria. Although not yet complete, new editions are continually being released.

- The Austrian Standard ÖNORM S 2101 (Hazardous Waste Index) lists the substances which qualify as hazardous wastes in Austria.
- Regulations of the Federal Ministry of Agriculture for the limitation of hazardous substances in sewage water and liquid wastes.
- The Austrian Standard ÖNORM M 6210 with regulations for the quality of liquid wastes from photographic laboratories and printing works.
- The Emission Control Act, BGBI. 559/1980, with regulations for the quality of fumes released from incineration facilities.

Laws pertaining to special waste are passed by the Federal State, but the nine individual provinces are responsible for the implementation of these laws. In the case of household waste, the laws are passed by the province and implemented by the communities. Therefore the Federal State is unable to uniformly regulate the treatment and disposal of waste throughout the country.

Currently new laws on waste management are being developed by several provinces. The new regulations will focus on the collection of hazardous wastes and secondary raw materials. For example, a 1986 amendment to the Salzburg Waste Law includes both the separate collection of household hazardous wastes by the communities and the mandatory participation in these collections by the public.

A new law on chemicals shall actualize the existing regulations. Some of its most important items will be obligatory labelling and declaration and the possibility of prohibitions, which will allow the authorities to control the traffic with certain chemicals.

In 1985, a study group with the chamber of workers and employees completed a draft of a Waste Avoidance Law. Its tasks shall be both the reduction and the influence on the composition of waste. The Federal Minister of Environment, Youth and Family shall be given extensive powers to control the introduction of new products. Debate on this draft in parliament is expected for 1988.

In 1987 an amendment of the Hazardous Waste Act 1983 was proposed by the Federal Ministry of Environment, Youth and Family with the aim to take into consideration the experience of the implementation of this law by the regional administrations of the nine provinces. It should also grant facilities to hazardous waste collectors and removers.

The Waste Oil Act 1979 has been amended by the Waste Oil Act 1986 (BGBI. 373/1986). In September 1987 the implementation decree for the Waste Oil Act BGBI. 383/1987 of the Ministry of Economic Affairs came into force. It regulates amounts and methods of measurement for the treatment of waste oil.

3

Canada

Surface: 7 237 778 km^2
Population: 25 354 000 inhabitants (based on 1986 census)
Political Structure:

- 261 municipalities between 10 000 and 50 000
- 55 municipalities between 50 000 and 100 000
- 35 municipalities over 100 000 (based on 1986 census)
- 14 Regional Municipal Governments whose constituents are municipalities included above.
- 10 Provinces:

Ontario	9 113 515
Quebec	6 540 276
British Columbia	2 889 207
Alberta	2 375 278
Manitoba	1 071 232
Saskatchewan	1 010 198
Nova Scotia	873 199
New Brunswick	710 422
Newfoundland	126 646
Prince Edward Island	568 349
Yukon	230 504
North West Territories	52 230

Waste Generation: 25 354 000 t/year total municipal and industrial/commercial, 2.7 kg/cap/day (average)
Industrial Waste: 60 000 000 t of which approximately 12 675 000 is included above and:
30 million • woodwaste (reused)
30 million • solid and liquid
• 10 million reused
• 3.2 million hazardous
Sewage sludge: 500 000 t of which 200 000 used
Agricultural: 40 million t
Mining tailings: 600 million t

The analysis and description of solid waste types are in Tables 1 and 2.

1.0 Waste Policy

The management of solid waste is primarily a municipal and regional government responsibility in Canada. Provincial governments set policies, regulate, give approvals, monitor operations, and provide financial incentives to implement policy. The federal government sponsors an annual Canadian conference on waste management and provides information exchange and support services, but has had little direct involvement in waste management. There is no preferred method or policy indicated to the municipalities by the federal and provincial governments with respect to landfill or incineration.

The methods for waste management vary across the country. The Eastern and Western areas have relied mainly on landfill due to availability of land, relatively low cost petroleum or hydroelectric energy, and lesser industrial and financial bases that result in lower priorities for higher cost recycling and incineration. A new Energy From Waste (EFW) incinerator is being built in the greater Vancouver regional district and there are some developments in recycling and waste exchange in Alberta and Saskatchewan.

In the province of Alberta, there is a beverage container "deposit depot" programme, whereby, the private sector has established approximately 200 depots where they will refund deposits for used containers.

"Waste exchanges" in several provinces provide a listing of waste materials and sources that are successful in attracting others who can use those materials and avoid need for disposal.

In central Canada, Quebec has EFW incinerators, in Montreal, and in Quebec City, with little recycling activity to date. In Ontario, four small incinerators are being built to service specific waste sources, and no large incinerators have been built. Separate recycling collections are developing in the five regions core area comprising a population of 3 623 200. Reduction, reuse and recycling are being advocated.

In the city of Toronto, a refuse fired steam plant to supply Canada's largest of four existing steam district heating systems, which is presently fuelled by natural gas, has been under active consideration for almost ten years, and has just reached the application stage for approval.

Metropolitan Toronto, a regional municipality in Ontario, providing solid waste management for a population of three million, is undertaking a Solid Waste Environmental Assessment Plan (SWEAP). This will be a comprehensive master plan to include reduction, reuse, and recycling, a specific landfill site and possibly the above-mentioned and other EFW plants and will establish the target amounts for each option.

Ontario has had abundant and low cost hydro- and nuclear electrical energy which may have deterred EFW incineration. The province is now encouraging both the electrical utility and energy regulators to ensure that alternative energy sources such as incinerator waste heat and biogas get due consideration. The Ontario hydroelectric power corporation is now required to buy electrical energy produced from waste at their displaced energy cost.

In Ontario, a policy described as the "4 R's" has been established consisting of:

Reduction: avoidance of waste by change in habit and lifestyle;

Reuse: in original state or purpose;

Recycling: reprocessing for use as a secondary material;

Recovery: landfill or incineration with energy recovery.

The "4R's" policy has been supported by a policy of provincial subsidy commencing at 50% and declining 10% per year to zero over a five-year period. Metropolitan Toronto supplements this with a contribution of 100% of net operating cost. For all other recycling initiatives, it provides a recycling credit equal to the avoided cost which currently is $15/tonne equal to 85% of the disposal fee.

These programs will be evaluated before five years to determine their cost and effectiveness in reducing reliance on conventional disposal options.

Canada's only biogas project is in operation in Kitchener, and the second is in preliminary design in Metropolitan Toronto in a 15.5 million t landfill site. Recently, EFW incineration has been developing slowly. Canada has been relying largely on landfill sites which are generally well controlled and many, such as the Keele Valley landfill site in Metropolitan Toronto, represent the "state of the art".

One indication of the impact of policy is the cost of service. Metropolitan Toronto, serving three million has a 1988 current budget including debt charges of $49.5 million, and a 1988 capital budget of $18 million. Included in the above are the costs of the Keele Valley landfill site which serves half of the population noted above, has been budgeted at $90 million for acquisition and capital development, and $13 million per year for the operating life of 15 years or 20 million t. These costs exclude collection costs.

2.0 Addresses

2.1 ISWA National member

Canadian Public Works Association:

Representative: R. G. Ferguson, Deputy Commissioner of Works, Municipality of Metropolitan Toronto, 439 University Avenue, 10th Floor, Toronto, Ontario, M5G 1YG

2.2 Government offices

Government of Canada
Environment Canada:
Minister, The Honorable Tom McMillan, 28th Floor, Les Terrasses de la Chaudiere, Hull, Quebec, K1A 0H3
Mr. J. Myslicki, Chief, Waste Management Division, Industrial Programmes Branch, Environment Canada, 351 St. Joseph Blvd., 13th Floor, Ottawa, Ontario, K1A 0H3 (Telephone 819 953 1390)
Province of Ontario:
Ministry of Environment:
Minister, The Honorable Jim Bradley, 135 St Clair Avenue West, 15th Floor, Toronto, Ontario, M4V 1P5
Acting Director, Waste Management Branch, Mr. Don Bartkiw, 40 St Clair Avenue West, Toronto, Ontario, M4V 1P5 (416 965 9668)
Other references can be obtained from the ISWA member.

2.3 Technical and professional associations

Canadian Public Works Association (CPWA), Suite 191, 253 College Street, Toronto, Ontario, M5T 1R5
Institute of Solid Waste, American Public Works Association, 1313 East 60th Street, Chicago, Illinois, 60637
Government Refuse Collection and Disposal Association, Mr. Frank Condlin, 10 Peel Centre Drive, Brampton, Ontario, L6T 4B9
National Solid Waste Management Association—Ontario Waste Haulers Association, 260 New Toronto Street, Toronto, Ontario, M8V 2E8

2.4 Research institutions

Ontario Research Foundation, Sheridan Park Research Community, 2395 Speakman Drive, Mississauga, Ontario, L5K 1B3

3.0 Information Transfer

3.1 Journals

Civic Magazine, 777 Bay Street, 5th Floor, Toronto, Ontario, M5C 2C8.
Proceedings of Annual Canadian Conference on Waste Management

3.2 Conferences

Canadian conference on waste management—October, by environment Canada, 12th Floor, Place Vincent Massey, Hull, Quebec, K1A 0H3.

3.3 University courses

Solid waste management, offered to graduates/undergraduate students, Professor G. Henry, and P. Byers, University of Toronto, 35 St George Street, Galbraith Bldg., Room 105, Toronto, Ontario, M5S 1A4

With reference to 2.4 and 3.3, the developments in design of landfill sites are largely decided by municipal staff, private consulting engineers, and the provincial regulating agencies. There is less involvement by universities and academics in research and design of landfills than in Europe.

4.0 Production and Composition of Waste

A typical description of solid waste types is shown in Table 1. Based on information from urban municipal operations, production of municipal waste in Canada is estimated at 12 677 t/year. This equals 0.5 t/cap/year, which would be municipally collected, or include sewage sludge, parks, waste, or other related municipal operations. The average daily generation is 2.7 kg/cap/day in total from municipal and nonmunicipal sources.

The composition of waste in Canada is not well-documented, perhaps because there is so little incineration and until recently, so little recycling from municipal sources that the information may not be considered necessary. A 1977 analysis performed for environment Canada is shown in Table 2.

5.0 Composition of Municipal Waste

Municipally collected waste represents approximately 50% of the waste received at municipal landfill and incinerator sites. The main features of Canadian municipal waste are the large percentages of waste newspapers, plastics, food and garden wastes. Newspapers are large in number and size of pages. In major urban areas, there are frequently two or three daily newspapers that weigh up to 4 kg per week in weight. This has given rise to separate waste newspaper collection in urban areas. The relatively high standard of living leads to the higher quantities of plastic packaging, short life novelty items and food waste. Large residential lot sizes result in more garden waste.

5.1 Composition of industrial/commercial waste

Industrial/commercial waste is assumed to be all waste in excess of that

Table 1 Description of solid waste types

Type of solid waste	Characteristics
Domestic Waste	• Household waste collected from households–residences and apartments • Light commercial waste collected by municipal forces
Municipal Solid Waste (MSW)	• Domestic waste from residences and apartment buildings • Street sweepings • Catchbasin cleanings • Sewage sludge • Ash • White goods (i.e. appliances) • Park wastes (i.e. leaves, brush, etc.) • Waste from municipal offices • Waste from libraries, boards, commissions etc. collected by municipal forces
Commercial Solid Waste	• Waste from restaurants, offices, commercial establishments, small industry • Waste from institutions (i.e. schools, churches, government offices)
Solid Nonhazardous Industrial	• Waste from Area Municipal/Metro construction projects • Private construction and demolition wastes • Foundry sands • Agricultural waste • Tyres • Industrial sludges (nonhazardous) • Nonputrescible waste • Wood, brush, etc. • Pallets, lumber, etc. • Off-specification or dated products • Manufacturing residues • Food processing wastes

Excerpt from Metropolitan Toronto SWEAP "Solid Waste Management Approaches", December 1987, by Proctor and Redfern and Senes Consultants.

collected municipally and described above, and may be delivered to municipal or private landfills or incinerators, approved for solid industrial and commercial waste. The quantity of industrial/commercial waste can only be estimated as the 50% balance of the total amount weighed at municipal disposal facilities. No analysis of composition is available, and quantity received at private sites may not be as accurately weighed and reported.

Table 2 Material/commodity analysis, Ontario (annual)

Component	As-generated Wet wt. (%)	As-generated S.E.*	As-generated Moisture content (%)	As-received Wet wt. (%)	As-received S.E.*	As-received Moisture content (%)
Paper	38.88	5.20	7.0	44.94	4.39	19.57
Kraft paper	9.51	3.04		10.75	3.13	18.48
Newsprint	9.33	3.03		10.61	2.91	19.30
Fine paper	7.91	2.97		8.07	2.74	11.03
Other paper	12.13	1.49		15.50	1.50	27.02
Glass	6.50	1.74	0.0	6.55	1.65	0.82
Beer containers	0.04	0.07		0.04	0.07	
Reuseable soft drink	0.23	0.15		0.23	0.15	
Nonreuseable soft drink	1.30	0.59		1.33	0.58	
Liquor and wine	1.50	0.54		1.53	0.53	
Containers—food	1.98	0.84		1.98	0.83	
Containers—other	0.28	0.24		0.30	0.24	
Flat and cullet	1.17	0.58		1.15	0.56	
Ferrous metals	5.34	0.82	0.0	5.49	0.76	2.88
Beer cans	0.00	0.00		0.00	0.00	
Soft drink cans	0.85	0.24		0.88	0.23	
Food cans	2.53	0.47		2.61	0.44	
Other	1.95	0.75		2.01	0.77	
Nonferrous metal	0.81	0.56	0.0	0.89	0.56	12.41
Aluminum	0.77	0.51		0.85	0.51	
Other	0.04	0.05		0.04	0.06	
Plastics	4.86	0.60	2.0	5.72	0.53	16.57
Containers	0.88	0.19		1.05	0.23	
Sheet film other	3.98	0.57		4.67	0.49	
Ceramics rubble	1.77	1.51	0.0	1.82	1.43	2.36
Lumber	3.36	1.84	15.0	3.36	1.77	13.76
Putrescible	30.55	6.33	70.0	22.59	4.59	56.63
Textiles, leather, rubber	3.55	0.72	4.0	4.11	0.98	14.54
Yard wastes, brush	3.30	2.41	50.0	3.29	2.34	31.55
Fines	0.77	0.34	2.0	0.93	0.38	20.09
Petroleum–chemical mix	0.31	0.20		0.31	0.20	
Total weighted moisture content (%)			26.52		—	25.18

* S.E. = 2S.D./$\sqrt{n}$ where S.D. is the standard deviation and n is the number of samples.
Excerpt from "Municipal Refuse Statistics for Canadian Communities of Over 100 000; Bird & Hale Ltd. for Environment Canada, 1977".

5.2 Collection methods

Urban areas receive edge of road municipal collection from one to three times per week including two regular collections and a Wednesday collection for bulky waste including garden waste. Oversize items are usually collected by arrangement on the bulk collection day. In order to make more efficient use of human resources and equipment, some municipalities have instituted waste newspaper collections on the bulk collection days (Wednesday), and others have authorized private sector collections biweekly. Three men crews are the norm in urban areas, but some are moving towards once per week collection and two or one man (driver only) crews.

Municipal collection is usually limited to 6–8 approved size and weight containers per collection day. The exact number of containers allowed and the frequency of collection varies among municipalities. Industrial and commercial properties with waste in excess of these limits must obtain private collection services. Approximately 50% of urban waste is picked up in 0.17 m^3 polyethylene "green" bags.

Municipal collection vehicles are mainly standard chassis with a 19.1 m^3 rear loading compaction unit that average 5 t with a full load. Two Canadian cities are using mechanized collection equipment in selected multiple unit residential and commercial areas on a trial basis.

In Southern Ontario, several municipalities serving a total of approximately 3.5 million for collection, are initiating separate multimedia curbside collections some using specialized multicompartment, noncompacting truck bodies. Initially three compartments will be used for waste newspapers, metal and mixed glass containers. Increasing sophistication and other designs may be required in future, if separate collections are extended to include polyethylene terephthalate and glass containers separately by a colour and a wet fraction.

6.0 Municipal Treatment and Disposal Methods

Recycling (Municipal)	
Average	0.25%
(Ontario only)	(1.7)
Composting	insignificant
Baling	none
Shredding	none
Incineration	
Average	4%
(Ontario only)	(4%)
Landfill	95.75% approx.

6.1 Incineration

It is estimated that 4% of Canada's solid waste is incinerated.

In order to investigate how incinerator emissions can be reduced, environment Canada established the National Incinerator Testing and Evaluation Program (NITEP). In an extensive four-year program, NITEP is testing the performance of the most popular EFW incinerator designs in Canada and assessing new air pollution control technologies that can be adapted to both new and existing incinerators.

The major incinerators are listed in Table 3. As indicated by the Table, interest in incineration is increasing, and several new EFW incinerators are under construction or planned. Regulation standards respecting emissions have not been raised to levels achievable with best available technology (BAT). Several provincial jurisdictions including Ontario can be expected to introduce more stringent regulation standards to BAT levels. Incineration as a disposal method declined in the 1970s and early 1980s. For example, Metropolitan Toronto closed seven old incinerators during that period and now maintains only one in operation, which handles 4% of its waste. In 1987, with land areas fully developed and faced with obtaining new landfill sites outside of its political jurisdiction, Metropolitan Toronto has undertaken a comprehensive Solid Waste Environmental Assessment Study (SWEAS) that will consider all options including one of at least 25% incineration in the near future.

6.2 Material recovery and composting

There has been no significant mechanized post collection solid waste processing in Canada to date. There has been some shredding, magnetic separation or composting. A small demonstration plant was operated for ten years in Metropolitan Toronto with shredding, corrugated picking, magnetic separation, air classification and composting. During this demonstration, refuse derived fuel (RDF) was test burnt in cement kilns on a trial basis only. No RDF is in use at this time.

Some preliminary design was done on refuse-derived fuel for coal substitution in the Metropolitan Toronto area. However, the project was not carried forward due to cost and reports of unsatisfactory shredding experience elsewhere. In Hamilton, Ontario, shredded refuse has been burnt in suspension for several years. (Refer to SWARU in Table 3.)

Efforts to recycle waste in the past have been concentrated on separate collection and depots for waste newspaper, and to a lesser extent glass and metal containers. While some newspaper programs are now 15 years old, they only exist in limited urban areas, mainly in Ontario. Public participation

Table 3 Selected examples of incinerators and facilities in Canada

Incinerator/EFW Location	Total design capacity (t/day)	Number of units	Incinerator type	Air pollution controls	Status
Quebec City (Quebec)	908	4	Mass burning—single stage	ESP	Operating
Vancouver (British Columbia)	800	2	Mass burn	Lime scrubber and baghouse	Nearing completion
P.E.I. Energy Corp. (Charlottetown, P.E.I.)	100	3	Two-stage controlled-air with waste fed by hydraulic ram	None	Operating
Victoria hospital (London, Ontario)	270	3	Two-stage controlled-air	Lime injection and baghouse	Nearing completion
Petro-Sun/SNC (Mississauga, Ontario)	360	4	Two-stage controlled-air	Lime injection and baghouse	Proposed
SWARU (Hamilton, Ontario)	600	2	Semi-suspension	Lime	Operating at 450 t/d
Commissioners Street (Metro)	900 rated 450 actual	3	Mass burn	ESP	Operating
Trintek Systems Inc. (Metro)	400	4	Multi-stage mass burn	Hydrated lime and baghouse	Proposed
Cherry Street RFSP (Metro)	1300	2	Single-stage mass burn	Hydrated lime and baghouse	Proposed
Pearson Airport (Metro)	40	1	—	—	—
Montreal	1200		Mass burn	ESP	Operating

Excerpt from Metropolitan Toronto SWEAP "Solid Waste Management Approaches", December 1987, by Proctor and Redfern and Senes Consultants.

Table 4 Metal criteria for fluid[a] anaerobically digested sewage sludges

Metal	Mean metal content of uncontaminated Ontario soils ($\mu g/g$)[b]	Maximum permissible metal content in soil ($\mu g/g$)[b]	Maximum permissible metal addition to soil (kg/ha)[d]	Minimum ammonium plus nitrate–nitrogen (NH_4^+–N plus NO_3^-–N) to metal ratios required in sewage sludge	Number of years to reach maximum recommended metal content in soil (column 3). based on columns 2, 3 and 5[c,d]	Minimum ammonium plus nitrate–nitrogen (NH_4^+–N plus NO_3^-–N) to metals ratios required to give maximum permissible metal content in soil (column 3) in 250 years[d]
Arsenic	7	14	14	100	50	480
Cadmium	0.8	1.6	1.6	500	30	4200
Cobalt	5	20	30	50	55	220
Chromium	15	120	210	6	45	32
Copper	25	100	150	10	55	45
Mercury	0.1	0.5	0.8	1500	45	8400
Molybdenum	2	4	4	180	25	1700
Nickel	16	32	32	40	45	210
Lead	15	60	90	15	50	75
Selenium	0.4	1.6	2.4	500	45	2800
Zinc	55	220	330	4	50	20

Excerpt from Ontario's guidelines for sewage sludge utilization on agricultural lands.

[a] Dewatered and dried sludges are to meet the column 5 criteria prior to dewatering and drying.

[b] Based on dry weight at 100°C. The terms $\mu g/g$ and mg/kg are interchangeable.

[c] Based on 135 kg of ammonium plus nitrate nitrogen per hectare per five years and sewage sludge having minimum ratios. Number of years is rounded off to the nearest five years.

[d] Columns 4, 6 and 7 apply to soils of mean content (column 2) and require adjustment for soils lower or higher in metal content.

Table 5 Metal criteria[a] for all aerobic sewage sludges and for all dried and dewatered anaerobic sewage sludges

Metal	Maximum permissible metal concentration (mg kg of solids)[b]
Arsenic	170
Cadmium	34
Cobalt	340
Chromium	2800
Copper	1700
Mercury	11
Molybdenum	94
Nickel	420
Lead	1100
Selenium	34
Zinc	4200

Calculations for aerobic sludge use on land

(1) To determine sludge acceptability, calculate "actual metal concentrations" and compare with the above permissible values

i.e. $$\frac{\text{Metal concentration (mg/l)} \times 10^6}{\text{Sludge solids concentration (mg/l)}} = \frac{\text{mg of metal}}{\text{kg of solids}}$$

(2) Calculate maximum application rate per five year period.

i.e. $$\frac{8 \times 10^6}{\text{Sludge solids concentration (mg/l)}} = \frac{\text{sludge (m}^3\text{)}}{\text{Land area in hectares}}$$

Excerpt from Ontario's guidelines for sewage sludge utilization on agricultural land.

[a] For permissible number of years to reach maximum metal concentrations in soils, refer to Table 2, column 6.

[b] Acceptability will be judged on the basis of the ratio of the average metal concentration during the preceding 12 months to the average concentration of solids during the same period or, at the discretion of local MOE staff and when solids and metals analyses are conducted once a month, on the basis of the last 3 results.

has been voluntary and having obtained only 20% of material available must be considered unsatisfactory. Waste newsprint deinking plants are now available. Trial programs in Ontario for curbside collection of glass and metal are expanding into separate source separated multimaterial curbside collection for newspaper, glass and metal containers. It is anticipated that this will remain the trend during the five years, during which these programs are subsidized, after which their public participation and cost effectiveness will be evaluated.

Land application of sewage sludge is permitted in compliance with guidelines. The Ontario guidelines are shown in Tables 4 and 5. In the major urban areas of Ontario sewage sludge solids are incinerated with waste heat recovery.

6.3 Sanitary landfill

Approximately 95% of solid waste is landfilled in Canada. Landfill design is "state of the art". Liners may be single or double, utilizing natural soils including recompacted clay, bentonite and augmented soil composites, and synthetic or flexible membranes or combination thereof. Leachate control under drainage and purge well backups are in use, and at the newest large site, Keele Valley landfill site, the flow from these purge wells, together with liner surface drainage will be recirculated through the landfill for rapid stabilization before the life of the liner expires. Methane recovery is operating at one small site in Ontario and under preliminary design for one 15.5 million t site in Metropolitan Toronto.

Typical landfill operations, based on the new Keele Valley Landfill Site serving 1.5 million people in Metropolitan Toronto area, are indicated above. These costs result in a disposal fee of approximately $20 per t to which a $30 per t reserve has recently been applied for 1988 to provide for expenditure anticipated for the potential future solid waste management options of recycling and incineration.

7.0 References

For further information with respect to overview contents and new installations, contact the Canadian ISWA representative, R. G. Ferguson, Deputy Commissioner, Metropolitan Works Department, 439 University Avenue, 10th Floor, Toronto, Ontario, M5G 1Y8 (416 392 8202).

7.1 Industrial waste

Little information is available on industrial waste composition.

7.2 Hazardous industrial wastes

Recent federal legislation requires registration and monitoring of chemicals and hazardous substances from production to disposal.

Some provinces have record systems which track the transportation of hazardous wastes from the source to the ultimate disposal point.

8.0 Legislation

Ontario will serve as an example because of the lack of direct federal involvement in solid waste, and the variations in provincial involvement. There are presently two pieces of provincial legislation which govern waste management in Ontario—the Environmental Assessment Act (EAA) and the Environmental Protection Act (EPA). The EAA applies to undertakings by the government and its agencies including municipalities. While certain public undertakings are exempt because their total cost falls below an established limit, most public projects having possible environmental implications are subject to the EAA.

The EPA also applies to private proponents. It focuses attention on the proposal and does not require all possible alternatives to be reviewed. The planning aspect and emphasis on relative merits of alternatives is the essential difference between the EAA and the EPA. Under both acts, existing environmental standards must be upheld.

The EAA was passed in 1975, and made applicable to municipalities in 1980. It provides an approval process through which activities and plans for activities of Ontario municipalities must pass before they are commenced or indeed before they receive any other legally required prior approvals or government financing. Its purpose is to cause proponents of undertakings to carry out early in a planning process a comprehensive study to determine the likely environmental and social effects of the proposed undertaking, to consider alternatives to it, as well as provide opportunities for the public to provide directional input and to comment on the undertaking, including whether or not it ought to proceed.

Prior to the EAA taking effect, for a waste disposal site to be approved and opened, a site-specific application was directed to the Ministry of the Environment. Prior to the ministry issuing a Certificate of Approval, a public hearing had to be held before the Environment Assessment Board (EA) (acting in an advisory capacity under the Environmental Protection Act) (EPA). The EA Board would then make a report back to the Ministry of the Environment which would have the final decision as to whether or not a Certificate of Approval would be issued and, if so, what terms and conditions ought to apply. The EPA hearing would concern itself with the basic safety, environmental and health aspects of the single proposed site and would not generally deal with the issue of alternatives.

Since 1980, with some limited exceptions, it is no longer sufficient for a municipality to pick a proposed waste disposal site and apply to the Ministry of the Environment for approval under the EPA. Rather, the EAA demands that proponents of a landfill, for example, consider alternatives to both the idea of landfill itself as well as alternative locations for the proposed landfill, taking into account potential impairment of the physical environment as well as impact on the social and cultural environments that may be affected.

Although a waste disposal site must still receive a certificate of approval under the EPA, the public hearing that would normally be required under that Act, will now normally be held under the EAA. In this case, the EA Board actually makes a decision as to whether or not the undertaking ought to be approved and under what terms and conditions. The EA Board's decision is, however, subject to variance by the Cabinet. Only one public hearing to satisfy the requirements of both Acts can be held with respect to the approval of a facility. The choice of whether the hearing is held under which Act is left to the discretion of the Minister of the Environment. However, it is likely, that if the undertaking falls within the scope of the EAA that the hearing would be held under that Act and not under the EPA.

Regulation 309 under the EPA (Ontario)

The waste management regulations under Ontario's EPA were substantially revised with particular emphasis on liquid industrial and hazardous waste and became effective on September 17, 1985. These regulations require generators, carriers and receivers of hazardous and liquid industrial waste (as specifically defined by the regulation) to register with or obtain approval from the Ontario Ministry of the Environment. Also, generators must provide information on facilities used for transportation and disposal, drivers transporting such materials to receive proper training, and additionally requires the use of prescribed forms so that such materials can be tracked from the point of generation to the point of disposal. (Insofar as the federal transportation of dangerous goods regulations require the keeping of records for shipments of such materials, the Ontario regulations attempt to avoid the duplication of effort.) The waste management regulations apply to waste that is originated, transported and imported into Ontario for disposal.

Many of the above references involve the province of Ontario and Metropolitan Toronto which has been selected as an example due to the difficulty in describing variations in policy and method across the country.

4

Denmark

Surface: 43 000 km^2 (excluding Greenland and the Faroe Islands)
Population: 5 111 108 inhabitants
Counties: 14 and Greater Copenhagen Area
Municipalities: 275
Cities (1984):
- Copenhagen 1 365 760 inhabitants
- 46 cities over 10 000 inhabitants
- 409 cities, 1000–9999 inhabitants
- 338 cities, 500–999 inhabitants
- 612 cities, 200–499 inhabitants
- Rural, 807 329 inhabitants.

1.0 Legislation, Policy and Administration

1.1 Legislation

The legislative framework for solid waste management is given by the following acts:

The Environmental Protection Act.
Law No. 372 of June 13, 1973; latest amendment by Law No. 392 of June 1987.

The Act on Waste Oil and Chemical Wastes.
Law No. 178 of 24 May 1972.

The Act on Chemical Waste and Landfills.
Law No. 262 of June 1983.

The Act on Recycling and Waste Reduction.
Law No. 297 first issued in June 1978; latest revision in May 1984 and now including the concept "cleaner technology".

For specific implementation of these laws, a series of government regulations are issued. Also of interest are certain action plans and special levies and programmes, e.g.

- "Dioxin action plan" of November 1985 and a follow-up document of May 1986 termed the "Recycling Action Programme".
- Waste charge of 40 Danish kroner/t levied on waste delivered at

incinerators and landfills to partly subsidize increased materials recycling.
- "Cleaner Technology Programme" initiated in 1987 to promote also low- and nonwaste technology.

To obtain a detailed list and further information, application should be made to the National Agency for Environmental Protection (NAEP), see Section 4.0 below.

1.2 Policy

Recycling (materials recovery) and cleaner technology now have the highest priorities in Danish waste management policy. Great efforts are made to change the production, collection and treatment of wastes; for instance, the government subsidizes new initiatives with the aim to change more traditional waste disposal systems into materials recovery systems or systems applying cleaner technologies.

The second highest priority is given to incineration since the possibilities of utilizing the energy produced are excellent with district heating in all larger towns in Denmark. The effort to minimize air pollution from incineration of wastes will in the next few years result in large investments in air pollution reduction and control systems.

The lowest priority is given to landfilling as a result of the problems with locating new sites for sanitary landfills in a rather densely-populated country with sensitive groundwater resources almost all over.

1.3 Administration

In general, environmental laws, by-laws, regulations and ordinances are administered through a hierarchical system starting at the top with the government and the NAEP and going downwards through counties and municipalities ending with private enterprises. Municipalities grant operating permission and exert control over private enterprises; counties similarly over municipalities.

Oil and chemical wastes are regulated through special legislation. These wastes are treated at "Kommunekemi", which is under direct supervision and control by the NAEP.

2.0 Categorization and Collection

2.1 Categories, amounts and composition

Where no year is indicated, 1985 is to be applied.

In Table 1, an account is made for Danish solid wastes (DSW) produced in trade, industry and households. Specifically included in Table 1 are:

Table 1 Annual waste production (from NAEP 1987, 1)

Constituent	t/year ($\times 10^3$)	% of total
Food	660	19
Paper & cardboard, dry	783	22
Paper & cardboard, wet	255	7
Plastic foil	86	3
Other plastics	57	2
Wood	97	3
Garden	325	9
Other combustible	164	5
Bottles	69	2
Other glass	58	2
Metal	439	13
Other noncombustible	439	13
Total	3432	100

- Wastes from production industry;
- Wastes produced by construction firms;
- Other trade wastes;
- Household wastes, including bulk items, garden waste and wastes from recreational and cottage areas.

Not included in Table 1 are:

- Recycled bottles, certain recovered metals, and paper and cardboard recovered directly in the paper and printing industry; these items are handled within existing well-functioning recycling systems; approx. 0.7×10^3 t/year.
- Wastes from construction, approx. 1.2×10^3 t/year.
- Asphalt from roads, approx. 0.4×10^3 t/year.
- Slags and ashes from power plants, approx. 1.5×10^3 t/year.
- Wastes from agriculture, forestry and fishery.

2.2 Collection, organization and technology

Organization

The collection of household waste is a municipal responsibility. However, most municipalities only involve themselves in the administration and the practical collection is carried out by private contractors to the municipality. A few large contracting companies operate regionally or even nationwide, but most contractors are small and operate only locally.

A few—mostly large—cities operate their own refuse collection department. Copenhagen and Aarhus, the two largest Danish cities, have long-term contracts with private, nonprofit utility companies.

All household waste collections are done from the yard or from container rooms inside the building and the empty receptacles are returned to the same location. Vehicles are generally manned by two or three men, all taking part in both the loading and the driving.

Weekly collection of household waste is almost universal. Only in the high density areas of the largest cities does more frequent collection occur.

Most collection schemes include a regular, e.g. monthly, or by-request collection of discarded furniture and other bulky waste from households. In many municipalities, even a garden waste collection is included or offered at extra cost.

Particularly in the large cities, an important proportion of the waste from offices, shops etc., is collected under the household collection scheme. However, the collection of commercial and industrial waste is generally left to private enterprises and carried out by direct contracting with the individual waste-producing companies. A few very large and many small locally-operating haulage contractors compete in a free market.

A pending change in the definition of municipal responsibility for the environment is likely to include the collection of commercial/industrial waste as from January 1, 1990.

Refuse receptacles

Since its introduction in the 1960s, the two-ply paper bag (100 l to 150 l) has held a remarkably high share of the market for household refuse receptacles. Only in the high-density areas of the largest cities, bins (100 l polyethylene) or wheeled containers (300 l/800 l fibreglass or polyethylene) are used in significant numbers.

In recent years, the modern two-wheeled containers (120 l, 200 l and 330 l) are gaining their share of the market and in some places replace the use of bags. Cheaper plastic bags or recycled polyethylene are also introduced in some cities.

Bags made from paper or plastic, are always placed in special stands with a lid and a protective metal net. For installation under refuse chutes, special equipment including automatic shifting devices is sometimes seen.

Large containers of the roll-on/roll-off type or skips of winch/rear loading type are mainly used for commercial/industrial waste and building waste but to a modest scale also for household waste from high-rise buildings and for garden waste or bulky household waste. For light industrial and commercial waste 300–800 l four-wheeled fibreglass containers are in widespread use.

Vehicles

In principle the use of paper bags allows the collection to be performed by simple flat-bed tipping trucks, and for many years small municipalities favoured the use of that sort of simple and cheap vehicle. The ever-increasing distances to regional disposal facilities have, however, resulted in a shift towards the use of modern compaction vehicles of the rear-loading type in order to reduce the number of trips and improve collection economy. Many different makes of both two- and three-axle compaction vehicles are in use for both household and commercial waste. Most vehicles are equipped with hydraulic container-lifting devices and some of the heavier vehicles have overhead-winches as well for the tipping of 4–10 m^3 rear loader skips.

A number of experimental vehicles for combined refuse collection and recycling have been tested and are still in use.

For the operation of a large variety of roll-on/roll-off containers ranging from 4 m^3 to about 30 m^3, traditional tipping frame vehicles of the cable/winch as well as the hook type are in widespread use.

3.0 Recycling, Treatment and Disposal

Handling of solid wastes is summarized in Table 2 by relative measures for certain waste categories.

Table 2 Waste handling in Denmark in 1985

Category	Recycling (%)	Incineration (%)	Composting (%)	Sanitary landfill (%)
Household	10	70	2	18
Bulk & garden	0[a]	34	0[a]	66
Commerce & industry	26	28	0	46
Construction	7	2	0	91
Energy production	67	0	0	33
Sewage sludge	30[b]	35	0	35

[a] Amount unknown.
[b] Application to land.

3.1 Recycling of materials

Three statutory orders, concerning municipal collection schemes from (a) private households, (b) private enterprises and (c) large-scale kitchens, were issued in 1986 by the NAEP. Under these orders, the municipalities shall by 1990 establish permanent collections of paper and glass from private households, paper and cardboard from private enterprises and food waste for animal fodder from large-scale kitchens.

The statutory orders constitute an important contribution to the development and spreading of recycling systems in the years to come. There are plans to introduce compulsory sorting of waste from private enterprises at all waste treatment plants, which receive more than 10 000 t of waste per year. As of 1988, all public institutions producing a certain amount of paper waste are obliged to collect their paper for recycling. The "Recycling Action Programme", cf. Section 1.1 above, includes a three-year examination programme on reduction of pollution from packaging material, and a three-year development programme.

In the long run, this development programme is meant to lead to a considerable reduction of waste through the introduction of so-called "green" systems and recycling of waste material from the building sector. The development programme also involves an investigation of the possibilities of establishing a production in Denmark of "grey" paper based on 100% waste paper. The programme emphasizes the importance of finding markets for goods based on recycling materials and development of sorting and production technology and of collection systems.

Paper, cardboard and wine bottles have for years been collected separately and used as raw materials in paper mills and glass works. The bottles are refilled after washing and the cullets sent to glass works. High rates of recycling have been obtained. Table 3 shows the recycling status for these materials in 1985.

Table 3 Recycling of paper, cardboard and glass. GENDAN (1986)

Material	Consumption ($\times 10^3$ t)	Recycling	
		$\times 10^3$ t	% of consumption
Paper/cardboard	997	312	31
Glass			
Cullets	—	25	—
Bottles	200–250 mill pieces	96 mill pieces	approx. 50

A number of central recycling plants for material recovery have been established in Denmark in recent years, cf. Table 4.

Table 4 Central recycling plants established in the 1980's, DAFOKA (1987)

Plant location	Type of plant	Input	Output
Vojens	PEL-recycling	Source separated PEL	PEL-granulate
Skive	Mechanical sorting	Mixed household, commercial and industrial wastes	RDF-pellets, paper, cardboard, plastic, metal
Mobile	Crushing mill	Construction wastes	Sand and gravel substitutes
Odense	Mechanical	Selected industrial & commercial wastes	RDF paper, cardboard
Roskilde	Manual/mechanical sorting	Selected industrial & commercial wastes	Paper, cardboard
Føvling	Manual and mechanical sorting	Selected wastes. Mixed household, commercial & industrial wastes	RDF, plastics, paper, cardboard, metals
Kolding	Waste to fodder factory	Food wastes from restaurants, etc.	Animal food

Wastes from energy production, especially power stations (fly ash, slag) have a high rate of recycling due to direct use in construction works (sand and gravel substitution) and land reclamation.

There seems to be a trend towards more systems based on source separation in combination with mechanical or manual sorting plants.

3.2 Biological waste treatment

Municipal, mixed wastes were composted to a large extent 30 years ago. However, compost utilization stopped because of high concentration of heavy metals that rendered the compost nonacceptable as fertilizer for food and fodder crops.

Separation of municipal wastes may involve a renaissance to composting. Noncontaminated food and garden wastes may be very suitable for treatment by composting or alternatively anaerobic digestion. Table 5 reflects such development for composting.

The total amount of composted waste amounts to less than 50 000 t/year; cf. also Table 2.

Table 5 Composting plants in operation

Plant location	Waste	Established
Frederikssund (M)	Mixed	1977
Frederikssund (M)	Separated	1986
Gentofte (P)	Garden	1986
Sengeløse (P)	Garden	1986
Odense (P)	Garden	1987
Aarhus (P)	Garden	1987

Where: P = Pile composting,
M = Mechanical, rotating drum.

3.3 Sludge

The total sludge production in Denmark is estimated to be 200 000 t/year dry matter. Approx. one-third of the sludge production is being incinerated. Ashes are then disposed of by landfilling. Direct landfilling of dewatered sludge is also practised. Though significant amounts, i.e. one-third of the total, are still disposed of this way, the trend is toward less landfilling of sludge. Land application of sludge, presently approx. one-third, is likely to increase, because the alternatives become less attractive in terms of technical and economical feasibility.

The present guidelines for sludge application to farmland NAEP (1984), recommends land application subject to cadmium less than 3 g/t dry matter and dosage of sludge according to crop requirement of phosphorus available in the sludge. Experience has shown the feasibility of obtaining cadmium less than 3 g/t dry matter in sludge, even in major cities, who restrictively grant permission for industrial effluents to sewers.

3.4 Incineration

Incineration is used widely, particularly for treatment of household wastes, cf. Table 2. Approx. 1.6×10^3 t/year of DSW are incinerated.

The total number of operating plants is 42 in 1987. District heating secures a high degree of energy recovery. Table 6 gives an overview. All plants are bulk-fired.

Twenty-seven of the plants are equipped with electrostatic precipitation for particulate removal. In 1986, the stack emission criteria were tightened,

Table 6 Incinerators in Denmark

Capacity t/h	<5	5–10	10–20	>30
Number of plants	19	16	5	2

so that most plants now have to improve both equipment operation and performance. Excerpts from the 1986 criteria are seen in Table 7.

As a consequence of the stricter regulations, the number of incinerators is expected to decrease, i.e. fewer and bigger plants will result.

Table 7 Incinerator stack emission criteria. More details in NAEP (1986)

Item of emission	Monthly average (mg/nm)	Annual average (mg/nm)
Particulates	40	—
HCl	100	—
Cadmium		100
Lead		1.4
Mercury		100
HF		2
SO		300
TOC		20
PCDD and PCDF	Control by operation, approx. measured as 2, 3, 7, 8 TCDD equivalents	1

3.5 Hazardous wastes

Industries, generating oil and chemical wastes (hazardous waste) must notify the local authorities and deliver the waste through a municipal collection system of 21 transfer stations to the central treatment plant—Kommunekemi—at Nyborg.

Kommunekemi started in 1975 and since then increasing amounts of wastes have been treated; cf. Table 8. Public concern about illegal disposal

Table 8 Treatment of hazardous wastes at Kommunekemi (t/year)

Type of waste	1975	1980	1985	1986
Waste oil	9229	17 466	35 056	40 244
Halogenated waste	1271	2679	2707	5266
Nonhalogenated solvents	2401	4160	5435	6357
Organic chemical waste	4115	22 463	31 706	36 306
Pesticide waste	—	172	1003	1174
Inorganic waste	251	6391	10 599	9835
Miscellaneous	783	2883	3903	4107
Total	18 050	56 214	89 409	103 289

has forced a stricter control on waste producers, but it is estimated that Kommunekemi still receives only two-thirds to three-quarters of the total hazardous waste generated.

At Kommunekemi, waste oil is heated for water separation and used as supply fuel for incineration of the organic wastes. Inorganic wastes are detoxified and the filtercakes are landfilled together with slags and flyashes from the two rotary kilns with modern gas cleaning systems. Kommunekemi is now using the full capacity of the plant and a third incineration line is expected within three years.

Approximately 5% of chemical wastes, e.g. waste solvents and silver are recovered by private enterprise. The technique for increasing this activity is available, but it is uneconomical due to low prices on virgin materials.

The municipalities can waive the delivery obligation, if other environmentally-acceptable handling be available. Especially aimed at the reduction of hazardous waste streams, which include significant amounts of hazardous waste from households, a development programme for cleaner technologies has been started by NAEP. In the years 1987–89, research and development regarding waste reduction technologies will be specially supported by government fundings, e.g. projects on substitution of hazardous raw materials (organic solvent to water base paints), process development in selected industries and product designs leading to more resource recovery and use of secondary raw materials.

3.6 Landfilling

Landfilling is used extensively, cf. Table 2, for several categories of waste. Approximately 3.5×10^3 t/year are landfilled.

A code of good practice has been established by NAEP (1982). Considerable work is going on to improve techniques, particularly regarding protection of groundwater, cf. RLLPG (1986).

The amounts of waste for landfilling may decrease significantly due to preventive measures of the society. However, there is a need for landfills in the future to dispose of e.g. noncombustibles and nonrecyclables.

4.0 Information, Addresses and Abbreviations

To obtain information on waste management, development and research in Denmark, there are several possibilities:

DAFOKA (Danish Committee on Solid Wastes, National member of ISWA)

Address: Ingeniorhuset
Vester Farimagsgade 29
DK-1606 Copenhagen V
Contact: Jeanne Møller
Telephone: 45 1 156565
Telefax: 45 1 937171

NAEP (National Agency for Environmental Protection)
Address: Strandgade 29
DK 1401 Copenhagen K
Telephone: 45 1 578310
Telefax: 45 1 572449
Telex: 31209 miljoe dk

ISWA Directory
This booklet contains addresses and names of all DAFOKA members, and identifies important professional institutions and persons dealing with solid waste research, development, administration, management and practical operations.

Within Denmark, national activities on solid wastes are reported in books, journals and at meetings, conferences and equipment shows. The language is Danish. Some examples can be given.

NAEP
Publishes several series of booklets and reports on new legislation, investigations, guidelines. Ask NAEP for a list of publications.

DAFOKA
Publishes a book series, e.g. reporting on results obtained by working groups. Ask DAFOKA for list of publications.
Arranges approx. ten one-day conferences per year.

MILJØ & TEKNOLOGI
Quarterly journal on environment and technology, covering especially solid wastes.
Published by: John Vabø A/S
Address: Svanemøllevej 34
DK-2100 Copenhagen Ø
Telephone: 45 1 298000
Telefax: 45 1 29 82 80
Telex: 19786 vaboe dk

DANMILJØ
Yearly trade fair for environmental technology covering water, air, waste, chemicals, noise, technology.
Exhibition as well as professional conferences and seminars.

Internationally, Danish research results are published in international journals, e.g. in *Waste Management & Research* by Academic Press, London, and normally in English.

5.0 References

GENDAN (1986) GENDAN-NYT, nr. 65 & 66, 1986 GENDAN, Copenhagen.

NAEP (1982) Vejledning i affaldsdeponering ("Guidelines for landfilling", in Danish), Vejledning nr. 4, 1982, NAEP, Copenhagen.

NAEP (1984) Slam fra rensningsanlæg ("Sludge from sewage treatment plants", in Danish) NAEP, Copenhagen.

NAEP (1986) Begrænsning af forurening fra affalds forbrænding ("Reduction of pollution from incinerators", in Danish), Vejledning nr. 3, 1986, NAEP, Copenhagen.

NAEP (1987, 1) Affaldsredegørelse ("Assessment of wastes", in Danish). Redegørelse nr. 1, NAEP, Copenhagen.

NAEP (1987, 2) Danish Waste Policy (in English). Redegørelse nr. 2, NAEP, Copenhagen.

RLLPG (1986) Reclamation of landfill leachate polluted groundwater, Programme proposal, Env. Eng. Dept., Technical University of Denmark, Lyngby.

5

Federal Republic of Germany

Surface: 248 717 km^2
Population: 61 000 000 inhabitants

1. Waste Policy

The federal government's environmental policy is characterized by the following three principles:

- The Principle of Prevention
- The Polluter-Pays Principle
- The Principle of Co-operation

Legislative competence in the field of waste management lies with the federal government, the executive competence with the individual "Länder" (federal states). The legislation provides a documentation of the state's responsibility for the environment with special emphasis being placed on the maintenance of a worthy environment for mankind, effective protection of nature, economic use of resources and elimination of damage which already exists. The main priority of the federal government is to avert danger and remove concrete causes of damage to the environment. In keeping with the principle of preventing damage to the environment wherever possible, current waste management policy gives priority to avoidance and reutilization of waste over waste disposal. Thus the new Waste Act of August 27, 1986 marks the transition from a policy concerned with waste disposal to one which emphasizes reutilization of waste. As well as this, the federal government has restructured its waste management policy to include the following aims:

- Improved monitoring of waste deposited before 1982,
- Improvement in quality of waste by reduction in the level of harmful substances and separate collection of waste containing harmful substances,
- Increased safety measures for disposal of hazardous waste through the introduction of technical regulations,
- Drastic improvement in the quality of waste by placing producers and distributors of products containing harmful substances under obligation to take back waste and through the introduction of regulations on packing,

- Placing more stringent requirements on the disposal of synthetic oils and waste oils which contain harmful substances.

The categorization of waste recovery, depositing of waste and connected measures under the main heading of "disposal" has automatically helped to create linguistic clarity; through this it has become possible to define the order of priority in disposal procedures—reutilization of waste gaining priority over depositing.

Thus the federal government is authorized to issue regulations to attain waste management goals which include the labelling of certain products, separate disposal, obligation to take back and return certain types of waste. A further important task is the preparation of "Technical instructions for waste" which will standardize the conditions for the permission and operation of waste disposal plants as well as their supervision. Regulations on waste oil have also been integrated into the Waste Act and the former Waste Oil Act rescinded. All waste oils are covered.

Utilization of waste is not, however, required at all costs. The obligation exists only where it is technically feasible and where a market for retrieved substances already exists or can be created. Comparison of cost of recovery and depositing does not, however, necessarily put potentially more expensive recovery procedures at a disadvantage, as the protection of such a limited resource as the countryside can only be attained at a price. The protection of the soil and ground water is the main objective in the waste utilization policy, the recovery of secondary raw materials a welcome by-product.

Liquid wastes also come under the regulations laid down by the Waste Act, as the concept of "sewage" has proven to be too narrow and in the future, no differentiation will have to be made between waste and sewage in decisive cases.

1.1 Ministry and departments

Under the decree of June 5, 1986 the federal chancellor instigated the establishment of a new Federal Ministry for the Environment, Conservation and Reactor Safety. The responsibility for the environmental policy of the federal government is concentrated within this ministry which makes it possible to deal effectively with environmental protection matters and nature conservation. The Federal Ministry for the Environment, Conservation and Reactor Safety is responsible for:

- Maintenance of air purity;
- Noise abatement;
- Protection of bodies of surface and ground water;

- Water resources and supply;
- Soil conservation;
- Waste management;
- Ecological protection;
- Species protection;
- Protection of biotopes and sensitive areas;
- Health aspects of environmental protection;
- Chemicals legislation;
- Toxic substances in food;
- Safety of nuclear power plants;
- Supervision of licensing, operation and disposal for nuclear plants;
- Legislation on radiation;
- Radiation hygiene;
- Cooperation with environmental protection organizations and international cooperation in all issues concerning environmental protection and reactor safety.

The ministry has supervisory power over the administration and technical work of the Federal Environmental Agency and the Federal Research Institute for Conservation and Ecology. It also participates in the supervision of the technical work of other federal institutes. With the commission for reactor safety, the radiation protection committee and the council of specialists on environmental matters the ministry has the expertise of reputable scientists at its disposal.

1.2 New legislation

Since 1986, a Waste Act has been in force which provides for new measures against toxic waste and ever-increasing quantities of refuse. The avoidance and recovery of residues and not disposal receive main emphasis under this act. State precautions prevent on the one hand, the depositing of hazardous substances in landfills and authorize on the other, intervention to ensure that quantities of waste for depositing be reduced. Just as individual citizens are required to make an effective contribution to environmental policy, technical solutions to bring about the stemming of waste accumulation are demanded of industry. Should the manufacturing industry not voluntarily create conditions for the attaining of a reduction in waste, in the area of packaging, for example, the Minister for the Environment has the power to impose certain conditions stipulating the use of packaging. This could mean, for example, that beverages would only then be sold in returnable bottles, disposable packages could only be used if a deposit were charged and completely nonreturnable packaging banned. It is difficult to reach and implement voluntary agreement, for example, between the Minister for the

Environment and the packaging and drinks industry as considerable conflicts of interest exist between the various branches of the industry.

Apart from this, the proper disposal of motor oil which becomes a waste product when oil changes are carried out in garages and filling stations or in industry, no longer comes under a special act for waste oil but has been incorporated into the Waste Act. However, attention should not be focused on the difficult commercial and technical problems involved in waste disposal but on opportunities which above all are open to medium-sized firms through the tasks set by waste management.

There is reason for some hope that the acceptance of responsibility on the part of individuals and industry along with the realization of the ecological necessity will in the future help to solve the urgent problems in the field of environmental protection.

2.0 Addresses

2.1 Federal offices

Der Bundesminister für Umwelt-, Naturschutz und Reaktorsicherheit	5300 Bonn 1 Adenauerallee 139–141 Tel. 02 28 56 1 Telex: 8 85 790
Umweltbundesamt President: Dr. Heinrich Freiherr von Lersner (Federal Environmental Agency)	1000 Berlin 33 Bismarckplatz 1 Tel. 0 30 8903–1 Telex: 1 83 756

Central institution is established to achieve a more effective collaboration among existing federal agencies and institutions working in the field of environmental research and to assume tasks below the ministerial level that fall within the competence of the federal government in the environmental sector. Interdisciplinary tasks performed include in particular:

- Development and maintenance of an information system for environmental planning and of a central system for the documentation on environmental data;
- Informing the public on environmental issues;
- Rendering of central services and assistance to the various departments for their research activities and for the coordination of the environmental research work of the federal authorities;
- Support in assessing the environmental impact of measures taken by the federal authorities;
- Checking of the general formulae of washing and cleansing agents as provided for in Article nine of the Detergents Act;

- Development of the scientific and technical bases for regulations and administrative rules in the sector of waste management;
- Checking of, and investigation into, procedures and installations in all sectors of waste management, in particular: waste disposal and utilization; recording, measuring and monitoring of facts and findings that are important in waste management;
- Techniques for avoiding wastes in the production of goods.

Umweltbundesamt Dokumentationszentrale Wasser (DZW) (Federal Environmental Agency–Central Documentation Office Water)	4000 Düsseldorf Rochusstr. 36

Documentation of, and information on, literature in the sector of waste management.

Bundesgesundheitsamt (Federal Health Office)	1000 Berlin 33 Thielallee 88–92

Scientific work for the Waste Water Charge Act, the Federal Water Act, the Epidemic Control Act, the Detergents Act, the Foodstuffs and Commodities Act, the Drinking Water Ordinance.

Institut für Wasser, Boden und Lufthygiene (Institute for Water, Soil and Hygiene)	1000 Berlin 33 Corrensplatz 1

Research work in the areas of sewage purification and protection of waters; measuring programme at the Rhine and Weser rivers. Water quality standards, assessment of water polluting substances. Investigation of drinking, bathing and waste waters. Analysis of waste water from nuclear power plants for radioactive substances. Investigation of the purification capacity of the subsoil of landfills. Hygienic aspects of waste disposal.

Bundesamt für Ernährung und Forstwirtschaft (Federal Office for Food and Forestry	6000 Frankfurt/Main Adickesallee 40

Problems of integrated water management (development of models).

Bundesanstalt für Gewässerkunde (Federal Institute of Hydrology)	5400 Koblenz Kaiserin-Augusta-Anlagen 15–17

Water cycle, water balance. Depth gauge measurements; water quality measurements (mass balance, self-purification), quality standards. Analytical and measuring procedures, model development. Status reports on water resources management. Waste water discharges and heat balance in coastal waters. Collaboration of the conventions for the prevention of marine pollution. Pollutant-load capacity of coastal waters.

Bundesforschungsanstalt für Naturschutz und Landschaftsökologie (Federal Research Institute for Nature Conservation and Landscape Ecology)	5300 Bonn 2
	Konstantinstr. 110
	Tel. 02 28 84 91 0
Chair: Dr. Walter Mrass	

Pollution of water bodies (bioindicators). River basins meriting protection.

Statistisches Bundesamt (Federal Statistical Office)	6200 Wiesbaden
	Gustav-Stresemann-Ring 11

Statistics on public water supply and public waste water management. Waste statistics.

Strahlenkommission beim Bundesminister für Umwelt, Naturschutz und Reaktorsicherheit	5000 Köln
	c/o GRS mbH
	Schwertnergasse 1
Director: Prof. Dr. Christian Stretter	Tel. 022 1 20681
Dir.: Dr. Detlef Gumbrecht	

2.2 Local authority groups

Verband Kommunaler Unternehmen e.V.	5000 Köln 51
	Brohler Str. 13
Pres: Dr. Manfred Rommel	Tel. 02 21 37 70 1
Dir: Dr. Karl Ahrens, MdB	Telex: 2 627 2 214 165

Verband Kommunaler Städtereinigungsbetriebe e.V.	5000 Köln 51
	PF: 51 06 20
Pres: Ltd. BauDir. Dipl.-Ing. Karl Ley	Lindenallee 13/17
Management: Dtsch. Stadtetag.	Tel. 02 21 37 71 2 80
Dipl.-Ing. Hans-Joachim Müller	Telex: 8 882 617

Verband Privater Städtereinigungsbetriebe e.V.—Bundesverband—Arbeitgeberverband	5000 Köln 90
	PF: 90 08 45
	Hauptstr. 305
Pres: Norbert Rothnmann	Tel. 022 03 810 75
Dir: StDir. a.D. RA Rudolf Trum	Telex: 8 874 571

Waste management

Länderarbeitsgemeinschaft Abfall Bay.	8000 München 81
StaMin. f. LdsEntwicklung u.	PF: 81 01 40
Umweltfragen	Rosenkavalierpl. 2

Chair: Ltd. MinR Dipl.-Ing. Fröbe Dir: RegR. Dr. Gerhard Rosenbauer	Tel. 089 92 14 32 43 Telex: 5 24 295
Arbeitsgemeinschaft für Abfallwirtschaft Chair: Beigeordneter Hans-Georg Lange Dir: Dipl.-Ing. Hans-Joachim Müller	5000 Köln 51 Lindenallee 13–17 Tel. 02 21 377 12 80 Telex: 8 882 617
Bundesverband Sonderabfallwirtschaft e.V. Chair: Dr. Klaus Hilger Dir: Dipl.-Vw. Rolf Eder	5300 Bonn 3 Siebengebirgsstr. 106 Tel. 02 28 48 00 25 26 Telex: 8 861 162

2.4 Protection of the environment

Deutsche Forschungsgemeinschaft Koordinator für Umweltforschung Dr. Albrecht Szillinsky	5300 Bonn 2 Kennedyallee 40 Tel. 02 28 8 85 1
Verein für Wasser-, Boden- u. Lufthygiene e.V. Chair: OStDir. a.D. Hans-Diether Imhoff Dir: Dipl.-Ing. Helmut Schönberg	1000 Berlin 33 Corrensplatz 1 Tel. 0 30 8 66 23 42
Arbeitsgemeinschaft ökologischer Forschungsinstitute Chair: Dr. Peter von Gizycki Dipl-Biol. Petra Günther Dipl.-Ing. Christoph Zimmermann	2800 Bremen 21 Waterbergstr. 21 Tel. 04 21 64 79 44
Institut für Energie und Umweltforschung Heidelberg e.V. Chair: Thilo Koch Dir: Barbara Steinhilber Schwab	6900 Heidelberg Im Sand 5 Tel. 0 62 21 1 01 01 or 1 29 56
Öko-Institut- Institut für angewandte Ökologie Chairman of the board: Günter Altner Dir: Dr. Rainer Greißhammer	7800 Freiburg Hindenburgstr. 20: Tel. 07 61 3 64 39
Deutsche Stiftung für Umweltpolitik Chair: Dr. Wolfgang E. Burhenne	5300 Bonn 1 Adenauerallee 214 Tel. 02 28 2 69 22 16 or 2 69 22 17

Institut für europäische Umweltpolitik Dir: Prof. Dr. Ernst Ulrich von Weizsäcker	5300 Bonn 1 Aloys-Schulte-Str. 6 Tel. 02 28 21 38 10 Telex: 8 86 885
TÜV Rheinland e.V. Institut für Materialprüfung und Chemie Dir: Dr.-Ing. Dieter Schlege	5000 Köln 91 Am Grauen Stein/ Konstantin-Wille-Str. 1 Tel. 02 21 832 93 0 Telex: 8 873 659 FAX: 83 93 18 13
Institut für Energietechnik und Umweltschutz Dir: Dr.-Ing. Eberhard Plaßmann	
Dachverband wissenschaftlicher Gesellschaften der Agrar-, Forst-, Ernährungs-, Veterinär- und Umweltforschung e.V. Pres: Prof. Dr. Wilhelm Henrichmeyer Dir: Dipl.-Agrarbauing. Peter Moog	6000 Frankfurt a.M. 1 DLG Zimmerweg 16 Tel. 0 69 71 68
Bundesverband der Deutschen Industrie e.V. Pres: Dr. Hans Joachim Langmann Dir: Dr. Siegfried Mann Press officer: Dipl.-Kfm. Dieter Rath	5000 Köln 51 PF: 51 05 48 Gustav-Heinemann-Ufer 84–88 Tel. 02 21 37 08 00 Telex: 8 882 601 Telex: 5 46 00

Organization of the 34 business associations formed to protect and promote the areas of common interest of the different branches of industry represented by their members, a total of some 100 000 firms.

Ausschuß für Umweltpolitik Bundesverband der Deutschen Industrie e.V. Chair: Dr. Otto Koch Dir: Dr. Eberhard Meiler	5000 Köln 51 PF: 51 05 48 Gustav-Heinemann-Ufer 84–86 Tel. 02 21 37 06 00
Thermo Prozess und Abfalltechnik Chair: Heinz Berger Dir: Dr. Arnold F. Lahrem	6000 Frankfurt a.M. 71 PF: 71 08 64 Lyoner Str. 18 Tel. 0 69 660 30 Telex: 4 11 321 4 13 152

Internationales Institut für Umwelt und Gesellschaft
Forschungsschwerpunkt Umweltpolitik
Dir: Prof. Dr. Ernst Simonis
1000 Berlin 30
Potsdamerstr. 58
Tel. 0 30 26 10 71

Bund für Umwelt und Naturschutz von Deutschland e.V.
Chair: Dipl.-Forstw. Hubert Weinzierl
Dir.: RA Lorenz Graf
5300 Bonn 1
in der Raste 2
Tel. 02 28 23 00 01
Telex: 8 869 7 53

Bundes v. Bürgerinitiativen Umweltschutz e.V.
Dir. Chair: Christa Reetz
Helmuth Wilhelm
Klaus Scherenberg
5300 Bonn 1
Friedrich-Ebert-Allee 120
Tel. 02 28 23 30 99

Official membership of 350 community action groups. Advisory service for approx. 1000 groups.

Deutsche Umwelt-Aktion Gemeinnütziger e.V.
Chair: Dr. Klaus Boisserée
Dir: Rolf Kurt Stein
4000 Düsseldorf 1
Heinrich-Heine-Allee 23
Tel. 02 11 13 13 22 131216

Fachverband Messen und Ausstellungen e.V.
Chair: Heiko Könicke
8500 Nürnberg 50
Messezentrum
Tel. 09 11 8 21 02
Telex: 6 22 080

3.0 Educational Courses

Educational courses in the field of environmental protection on offer at universities and institutes of higher education:

GUIDE TO STUDIES IN ENVIRONMENTAL PROTECTION
3rd Edition, October 1982
vailable from: Federal Environmental Agency

Additional study programme:
Fachhochschule Nordostniedersachsen Fachbereich Bauingenieurwesen (Wasserwirtschaft und Kulturtechnik)
Herbert-Meyer-Straße 7
3113 Suderburg 1
Tel. 05829 551

4.0 Quantity and Composition of Solid Waste

4.1 Volume of waste and trends

Considerable efforts on the part of the federal states and local authorities have led to over 90% of households being integrated into the public waste disposal network with 100% participation in conurbations. Those in possession of household or similar types of waste are obliged by law to place it at the disposal of the bodies responsible for its disposal or third parties appointed by them. In 1983, the volume of waste to be collected, transported, treated and disposed of in the Federal Republic was ascertained to be 32.5 million tonnes, of which 26 million consisted of household waste and similar industrial waste and 6.5 million tonnes of bulky waste, street refuse and refuse from markets. The waste was disposed of in 3150 local authority plants. Besides landfills, 45 waste incineration plants and 19 composting plants were in operation. In 26 waste incineration plants, separation of used-substances was undertaken. The costs involved in disposal are very high. It is estimated that collection and transport costs are about 50–100 DM/t. 70% of this figure can be accounted for by personnel costs, depositing costs approximately 15–70 DM/t and incineration costs approximately 50–200 DM/t. This shows that full exploitation of avoidance potential in the area of household waste and similar industrial waste could lead to considerable savings for the national economy and individual firms. It has been shown that the composition of waste is constantly changing and that it presents an increasingly dangerous burden on the environment. In terms of number and capacity, facilities for safe waste disposal have perceptibly increased in past years but they still fail to meet the demand.

4.2 Composition and disposal of solid waste

In a nationwide analysis of household waste in 1979/80, the proportions in the various groups of substances were determined as reported in Table 1.

In a survey undertaken in 1985 by the "Verband privater Städtereinigungsbetriebe" (VPS) (organization of private urban waste disposal firms), whose members are responsible for the disposal of approximately 50% of household waste, 70% of wastes from industry and commerce similar to household waste and 100% of hazardous waste, the quantities of raw materials recovered from waste which could be reutilized in production processes were ascertained, obtaining the data reported in Table 2.

Apart from this numerous conclusions could be drawn: "Bring systems" (e.g. containers at depots) are less costly. The quantities recovered are however low. Collection systems (e.g. containers for collecting used sub-

Table 1 Components of solid waste and relevant proportions

Substances	Million tonnes	Percentage (%)
Organic	9.77	42.3
Paper/cardboard	4.62	20.0
Glass	2.68	11.6
Fine waste/minerals	2.66	11.5
Plastics, textiles	1.76	7.6
Iron, nonferrous metals	0.9	3.9
Wood, leather, rubber	0.53	2.3
Others	0.18	0.8
Total	23.1	100

Table 2 Quantities of raw materials recovered from waste which could be reutilized in production processes

	t		
Raw material	1983	1984	% increase
Glass	271 176	379 260	39.9
Paper	457 703	617 828	35.0
Ferrous metals	165 345	309 817	87.4
Nonferrous metals	4576	17 635	285.4
Plastics	25 205	33 502	32.9

stances) are more expensive than "bring systems" but they recover larger quantities. In each collection form, manual or automated sorting is necessary. High volumes of recovered materials bring about a perceptible reduction in waste. The recovery of secondary raw materials from household waste necessitates complicated sorting processes which have to be taken into account when fixing prices both for disposal and raw materials. Nevertheless, reutilization efforts will have to increase in the future, namely in those cases where natural resources are threatened and where the possibilities of extending existing disposal facilities or finding alternative sites are limited. In 1982, a capacity of less than six years was estimated for 1615 of the 3062 household-waste landfills in the Federal Republic of Germany.

Considerable quantities of waste which require special disposal treatment arise from commercial production processes and also as by-products of environmental protection measures such as maintenance of water and air quality (e.g. sewage sludge, dust from filters, slag, gypsum). The annual volume of hazardous waste lies in the order of four millions tonnes. Increases are to be expected due to the more careful separation of dangerous waste substances and to intensified environmental protection measures made possible by the development of improved environmental technology. This often means that problems connected with harmful residues are transferred from the field of water and air quality control to the waste management. In 1983, a total of 26 hazardous waste landfills, 18 waste incineration plants and seven chemical-physical plants were in operation.

Particularly problematic is the limited availability of suitable space in landfills or special plants for incineration and other treatment processes. At present the volume of space available nationwide for depositing hazardous waste—given that volumes produced remain constant—can be utilized for 10–14 years; some landfills have, however, already reached the limits of their capacity. These problems cannot be solved by transportation of waste to other countries. Thus, in the long term, acceptable solutions for safe disposal within the Federal Republic of Germany must be found. Particularly in the case of disposal of hazardous waste with the special technical requirements it entails, it is essential for economic reasons alone that plans be drawn up for the optimum location and operation of plants which should deal with as large a catchment area as possible.

5.0 Collection and Separation of Useful Material from Household Waste

Waste utilization is seen as an important and future-orientated disposal concept which gives priority to the utilization of substances retrieved from waste over depositing in landfills. Useful substances from household waste are usually separated by the following two procedures.

- Conventional collection and transport methods with processing of the untreated waste in centralized or decentralized sorting plants;
- Preliminary sorting of the untreated waste in the household or firm into required categories of substances.

In the former conventional procedure, the household waste is treated mechanically; the useful substances are selected manually, divided into different filter grades and undergo special sorting procedures e.g. magnetic separation. Paper and cardboard, tin, paper/plastic material which can be used for generating energy and raw materials suitable for composting can be sorted without difficulty. A particular advantage of this system is that the collection and transport procedure which already exists need not be altered. The outcome of separation is dependent on the separation procedure and conditions and the quality of the waste but not on the cooperation of the public. With careful plant operation, quality products suitable for utilization can be produced. At 50–70 DM/t separation costs are relatively low. Combination with other treatment procedures causes no major difficulties.

In comparison to mechanical waste treatment, the separate collection at source of useful materials places greater organizational demands on the collection procedures, as it can either be partially or fully integrated into the normal refuse collection routine or complement this with "bring" or "collect" systems. At present, various collection and sorting methods are in use in the "collect" and "bring" systems. There are differing opinions as to their advantages and disadvantages in terms of quantity of substances retrieved and quality and cost of the useful materials. The most common systems are:

- Door-to-door collection of paper, cardboard and textiles. These collections are usually organized by charity organizations or scrap dealers;
- Glass/paper containers with a capacity of 1.5–2.6 m^3 with the recent addition of noise insulation;
- Tin containers which are similar to those for glass collection and often placed beside these. This is practised in few cities as recovery quotas are low. Due to the low weight of the scrap, the relation between cost and outcome is decidedly unfavourable;
- Multicomponent containers which unlike the used glass collection containers are exchanged rather than emptied. They are subdivided into compartments for glass, paper and cardboard, tin, batteries or old medicines. The containers are usually driven away when at least one compartment is full and then emptied in centres for collection of used materials. The quality of substances collected is similar to that collected in the other container types. The container is removed when at least one compartment is full and as, in general, the compartments are filled at varying rates, the container volume is on the whole inefficiently used. Costs rise as, in principle, it is mainly the weight of the container which is being transported rather than that of the materials.

6.0 Disposal of Household Waste

6.1 Utilization of household waste

The new Waste Act creates an improved legal basis for the development of a modern, environmentally acceptable approach to waste management. The utilization of waste in terms of material or energy production attains, insofar as it is technically feasible and economically practicable, priority over the conventional approach of waste disposal. The economic viability of waste recovery is no longer judged alone against the costs of traditional depositing of waste in landfills but also takes into account the factor of protection of natural resources. In recent years, measures for waste recovery which aim at the reduction of quantities for depositing, recovery of raw materials and the protection of natural resources have been widely introduced. Collection facilities for used glass and used paper are available almost everywhere.

Thus, in the meantime, almost a quarter of Germany's used paper—4.5 million t/year (1986)—is collected from households, an increase in the separate collection of used paper is considered to be in the interest of local authorities as it reduces the burden on scarce and expensive landfill space. The recovery of used glass has also established itself through active public participation. In 1980, merely 20% of used glass (500 000 t) was reutilized. This has meanwhile risen to more than 40%—1.1 millions tonnes. It is the aim of the federal government to raise this to 1.7 million tonnes by 1990 with the development of efficient collection systems.

Tin and aluminium cans are also suitable for recovery. 240 000 t of tinplate from local authority waste incineration plants are reutilized in production processes. The generation of energy from plastic waste contributes to energy-saving in waste incineration plants. At present approximately 500 000 t of heating oil are being saved per year.

Waste-containing harmful substances (e.g. batteries, medicines, paints, solvents) is also increasingly being collected separately or suitable collection facilities are being made available to rid household waste of harmful substances. This has been common practice in industry.

One of the most important aims in waste management is the improvement in performance levels in waste disposal technology to meet those in production technology so that in the future only products which can more easily be reutilized will be marketed and which beyond this accommodate the interaction of waste disposal, trade in raw materials, service and production industries.

6.2 Depositing of household waste in landfills

Although the number of efficient, modern chemical-physical treatment plants with high qualitative and quantitative disposal capacities has on the whole increased, it has still not proved possible, despite considerable efforts by the local corporations, to supply enough landfills which are environmentally harmless. Thus the disposal of waste, and in particular of hazardous waste, in densely-populated areas is a growing problem, as waste which cannot be reutilized or put to thermal use has to be deposited in landfills. A factor which has added to this scarcity of space is the general opposition put up by local people which hinders the establishment of new landfills. Thus, the length of time which lapses between drawing up plans and actual operation cannot be calculated.

According to information currently available, the majority of landfills and rubbish tips presently in use—as well as those which have already been closed down—are dangerous if in continued use. A large number of abandoned and closed-down landfills in the Federal Republic have been identified where harmful residues and toxic substances are contaminating the soil and are responsible for polluting ground and surface water. Over the next ten years financing in the region of 17 thousand million marks is estimated as required for decontaminating the approximately 5400 tips which urgently need redevelopment.

The responsibility for administration and finance lies with the individual federal states. However, the federal states are only required to make financing available when those responsible for the damage can no longer be found or implicated. In such cases, joint solutions are at present common which involve financial cooperation between the federal states and the local authorities on the one hand and commercial associations on the other. The redevelopment of these abandoned, environmentally hazardous landfills is a potentially interesting field for commercial firms and represent a future challenge for innovative building companies.

6.3 Incineration of household wastes

The lack of sites for landfills has resulted in the necessity of intensified development of incineration capacities as well as other measures. At present, approximately 46 waste incineration plants are in operation, to which a further 23 are to be added in the next few years. Up until a few years ago, waste incineration, with its advantages of reduction in volume, generation of energy and the resulting product of easily deposited slag, seemed to present the solution to the problem of waste disposal. Thus, of the 25 million tonnes of household and similar commercial waste which have to be disposed of annually in the Federal Republic of Germany approximately one-third is incinerated, 3% converted to compost and almost two-thirds are deposited in landfills.

Waste avoidance has meanwhile taken priority over disposal, but it will not be possible to eliminate the need for incineration in the future. Noxious emissions of substances such as carbon monoxide, hydrochloric acid, sulphur dioxide, nitrogen oxides, heavy metals, dioxines and furanes have attracted criticism of waste incineration plants. It is essential that the tried and tested incineration techniques are adapted so that slag, filter ash, effluent and residues from flue gas purification can be reutilized or disposed of in an environmentally-safe fashion. A large number of incineration technology measures and techniques for the purification of waste gas is meanwhile available which enables threshold limits values to be adhered to in a way that is economically and environmentally acceptable. Thus small plants for the incineration of household and similar residues with a few megawatts capacity are increasingly gaining in importance and are characterized by consistent and complete incineration and low levels of pollutants in the flue gas resulting from the use of modern procedures which allow easy regulation of the incineration process.

Thus, if care is taken that waste incineration plants are equipped in keeping with latest technological developments and also that flue gas purification reaches highest standards, current knowledge implies that there can be no objection to the increased disposal of household and similar waste in waste incineration plants so that the burden on landfills will be reduced. Technological developments can meet the increased standards required and can also bring the dioxine problem under control.

6.4 Investment in environmental protection and financial support

Environmentally-safe disposal of waste will be increasingly dependent on the extent to which substances harmful to the environment used in chemical-technical production can be replaced by environmentally compatible ones. This implies a high level of investment. On the other hand, preventive environmental protection and also repair of existing damage is frequently seen as a "basic responsibility of industry".

Thus in 1986, German industry invested more than twice the amount in environmental protection as in 1980 in order to ease the burden on soil, water and air, as can be seen in Table 3.

Table 3 Environmental protection spending in the manufacturing industry (in thousand million DM)

Year	Investment	Operating costs	Total
1971	2	3.3	5.3
1975	2.5	4.5	7
1980	2.7	5.4	8.1
1986	7.5	8.9	16.4
Total 1971–1986:	50.4	86.8	137.2

Source: Statistisches Bundesamt, BDI.

For the financing of environmental investments, low-interest loans with favourable repayment periods are made available from the Marshall plan fund to local corporations as well as to commercial firms. In 1986, loans amounting to 420 million DM for the construction and extension of plants for waste disposal and waste recovery were sanctioned by the "Kreditanstalt für Wiederaufbau" in Frankfurt and the "Lastenausgleichsbank" in Bonn. In 1984, for example, funds for environmental protection amounted to 556 million DM which was approximately 15.3% of the total credit sanctioned.

Apart from this, in certain cases, the redevelopment of abandoned landfills and disposal sites is financed directly by federal funds, such as the fund for the "improvement of regional economic structure" as part of a programme to redevelop abandoned industrial sites or on the basis of an act to promote urban development, which includes urban renewal schemes under which derelict industrial sites are used for new purposes.

7.0 Disposal of Hazardous Waste

The local authorities as corporations responsible for waste disposal have concentrated largely on their recovery programme of used materials on glass and paper. However the separation of plastics and used metals, which account for one million tonnes of waste per year, has begun to gain in importance. As paper and glass do not pose a threat to the environment, in many areas it is ecologically suspect products such as batteries, medicines, solvent residues, paints and chemicals which are increasingly being collected separately and where possible reintegrated into production processes. Household waste contains approximately 12 000 tonnes per year of these undesirable substances. However, 15% of the used substances in household waste cannot be recycled. These problematic substances which cannot be reutilized have to be deposited along with filter residues from local authority waste incineration plants in 2800 landfills including underground landfills. Due to success in purification of air and water, redevelopment of contaminated sites and general progress in environmental technology, the quantity of hazardous waste produced is constantly increasing.

Reliable technology has been developed for the treatment and disposal of hazardous waste from trade, industry and hospitals. Besides depositing techniques and chemical-physical treatment, waste incineration has attained increasing importance in this field. If properly incinerated, the conversion of waste into inert material which is suitable for long term depositing can be guaranteed, as persistent organic toxic substances are destroyed and can no longer harm the environment. Thus, given the current level of technical development, thermal treatment of hazardous waste is a method of disposal which offers a high level of environmental compatibility even when all possible environmental protection aspects are taken into account.

The construction of new hazardous waste incineration plants is a major prerequisite for overcoming this problem which at present is solved most unsatisfactorily by "exporting" it to neighbouring countries. Surveys have shown that the Federal Republic needs at least ten new plants with throughput capacities of approximately 60 000 t/year. Construction costs for a hazardous waste incineration plant with an annual capacity of 50 to 60 000 t and equipped with the latest technology are about 200 million DM. At present, some 16 hazardous waste incineration plants are at various stages in planning; none however are currently under construction.

Principally planned for the thermal treatment of hazardous waste, are rotary furnaces which are suited to the disposal of many problematic waste substances. Pyrolysis techniques are, however, also being tested for plastics wastes and old tyres. The types of waste for which new plants should be built have not yet been established in detail. This will take place at an advanced stage in planning. Basically all combustible toxic substances can be treated. Further thermal processes (e.g. fluidized bed furnaces and arc furnaces) are not in commercial use but, as they have not been sufficiently tested, are used only in special cases. Similarly incineration involving various types of grate firing systems, multistage grate and muffle furnaces can only be considered for limited use in special cases. For most types of hazardous waste, thermal treatment will become an indispensable method of disposal so that environmentally compatible disposal of unavoidable and unusable harmful substances can be guaranteed.

A further aim is also to avoid creating problems in the future from sites contaminated by hazardous waste.

6
France

Surface:	550 000 km^2	
Population:	54 330 300 inhabitants	
Municipalities:	36 453	
Population breakdown:		
Over	300 000 inhabitants	5
Between	100 and 300 000 inhabitants	31
Between	10 000 and 100 000 inhabitants	752
Under	10 000 inhabitants	35 665

1.0 General Remarks on Solid Waste Policy

The solid wastes policy is divided into three different lines of action: firstly, action concerning manufacturing processes and consumption habits with a view to decreasing the amount of waste produced at the source; secondly, action to increase recovery of certain materials contained in the waste; and lastly, action to dispose of the nonrecoverable matter in conditions that are satisfactory with regard to the environment.

The legal and regulatory framework presently governing the disposal of solid waste is based on two main texts, i.e. the laws of July 15, 1975 on waste disposal and the recovery of materials; the law of July 19, 1976 concerning installations classified for the purpose of environmental protection to which must be added, particularly with regard to industrial waste, regulations on the transport of hazardous substances (RTHS); plus the law of July 12, 1977 on the control of chemicals (as amended by the law of October 21, 1982).

The law of July 15, 1975 relating to the evacuation of solid waste and the recovery of certain materials constitutes a basic text that makes the persons producing the waste responsible for finding and implementing a satisfactory solution for its disposal. It puts municipalities in charge of disposing of household refuse. Those who produce toxic or dangerous waste must be capable of supplying full information on the conditions in which it is produced, collected, transported and processed. The law also provides the possibility of passing contracts with the industrial sectors concerned in order to develop the reuse of salvaged materials. Lastly, industrialists must make sure that the products they manufacture can be easily recovered and disposed of when they are no longer serviceable.

The body in charge of coordinating and activating this policy is the Ministry of the Environment (Direction of Water, Prevention of Pollution and Hazards; Solid Wastes Department), with the support of an inter-ministerial commission.

The regional departments of research and industry in charge of controlling classified installations (whether industrial facilities producing waste or plant treating industrial and household waste) enforce the rules, under the authority of Government Representatives.

The Basin Authorities, by their system of financial assistance, also take a hand in encouraging the proper treatment of industrial waste likely to pollute the water resources.

The national agency for the recovery and disposal of solid waste (in French ANRED), a public institution of an industrial and commercial character, under the supervision of ministers in charge of the environment and industry, is instructed to implement the official policy all over France.

2.0 Addresses

2.1 National correspondent of ISWA

Association Générale des Hygiènistes et Techniciens Municipaux
9, rue de Phalsbourg – 75017 Paris
Tel. 1 42 27 38 91

2.2 Ministries

Ministère de l'Equipement, du logement, de l'Aménagement du Territoire et des Transports
Services du Ministre Délégué, chargé de l'Environnement –
Direction de l'Eau et de la Prévention des Pollutions et des Risques (Service des Déchets)
14 Boulevard du Général Leclerc – 92524 NEUILLY
Tel. 1 47 58 12 12

Ministère de l'Industrie, des P & T et du Tourisme
101 rue du Grenelle – 75007 PARIS
Tel. 1 45 56 36 36

Ministère de l'Intérieur
Direction Générale des Collectivités Locales
4–12 rue d'Aguesseau – 75800 PARIS
Tel. 1 42 61 51 40
& 42 66 25 50

Ministère de l'Agriculture
Direction de l'Espace Rural et de la Forêt
78 rue de Varenne – 75700 PARIS
Tel. 45 55 95 50

Direction de la Qualité
44 Boulevard de Grenelle – 75732 PARIS Cedex 15
Tel. 45 75 62 25

Ministère des Affaires Sociales et de l'Emploi Direction Générale de la Santé (sous-direction de la prévention générale et de l'environnement)
1, place Fontenoy – 75700 PARIS
Tel. 45 67 55 44

2.3 National agency

- Agence Nationale pour la Récupération et l'Elimination des déchets (ANRED)
  2, Square Lafayette – BP 406
  49004 ANGERS Cedex
  Tel. 41 87 29 24

Basin Authorities (Agences Financières de Bassin)

- Adour-Garonne
  84 rue du Férétra – 31078 TOULOUSE Cedex
  Tel. 61 25 21 51

- Artois-Picardie
  764, Bd Lahure – 59508 DOUAI
  Tel. 27 87 01 94

- Loire-Bretagne
  Avenue Buffon – BP 6339 – 45063 ORLEANS Cedex
  Tel. 38 63 08 16

- Rhin-Meuse
  Le Longeau – BP 36 – 57160 MOULINS-lès-METZ, ROZERIEÜLLES
  Tel. 87 60 48 88

- Seine-Normandie
  51, Bd Salvador Allende – 92027 NANTERRE Cedex
  Tel. 47 76 44 24

- Rhône-Méditerranée-Corse
  31, rue Jules Guesde – Pierre Bénite – 69310 LYON
  Tel. 78 50 16 40

- Agence Française pour la Maîtrise de l'Energie
  27 rue Louis Vicat – 75015 PARIS
  Tel. 46 45 44 71

- Agence pour la Qualité de l'Air
  Tour GAN – Cedex 13 – 92082 PARIS LA DEFENSE
  Tel. 47 76 44 21

2.4 Main research institutions

- Bureau de Recherches Géologiques et Minières (BRGM)
  BP 6009 – 45060 ORLEANS Cedex
  Tel. 38 63 80 01
  191, rue de Vaugirard – 75737 PARIS Cedex 15

- Institut de Recherches Hydrologiques (IRH)
  10 rue Ernest Bichat – 54000 NANCY
  Tel. 83 96 65 10

- Institut National des Sciences Appliquées de LYON (INSA)
  20 Avenue Albert Einstein – 69621 VILLEURBANNE Cedex
  Tel. 78 94 81 12

- Laboratoire Central des Ponts et Chaussées
  58 Bd. Lefebvre – 75732 PARIS Cedex 15
  Tel. 45 32 31 79

- Centre d'Etudes et Recherches des Charbonnages de France (CERCHAR) – BP 27 – 60103 CRETEIL
  Tel. 44 55 35 00

- Institut National de la Recherche Agronomique (INRA)
  149 rue de Grenelle – 75007 PARIS
  Tel. 45 51 41 09

- Institut de Recherche de Chimie Appliquée (IRCHA)
  BP 1 – 91710 VERT-le-PETIT
  Tel. 64 98 24 75

- Centre d'Experimentation du Machinisme Agricole du Génie Rural des Eaux et des Forêts (CEMAGREF)
  13 rue du Cucillé – BP 1312 – 35016 RENNES Cedex

Research in the waste field is namely coordinated by the "Ground and Wastes Committee" which is located by the Research Department of the Environment Ministry.

2.5 Professional unions

- Confédération Nationale du Déchet (C.N.D.) et Fédération Nationale des Activités du Déchets (FNAD)
  72 rue d'Amsterdam – 75009 PARIS
  Tel. 42 85 26 30

- Syndicat National des Industries du Déchet (SNID)
  10 rue de Washington – 75008 PARIS
  Tel. 45 63 70 40

- Union Nationale des Exploitants de Décharges (U.N.E.D.)
  3 rue Alfred Roll – 75017 PARIS
  Tel. 47 66 03 64

- Fédération Nationale des Syndicats des Industries et Commerces de la Récupération (FEDEREC)
  14 rue de Courcelles – 75008 PARIS
  Tel. 43 59 64 03

- Chambre Syndicale des Entreprises d'Enlèvement de Déchets Industriels (E.D.I.)
  72 rue d'Amsterdam – 75009 PARIS
  Tel. 42 85 26 30

- Chambre Syndicale Nationale des Entreprises de Services Auxiliaires des Collectivités et Administrations Publiques (TACAP)
  72 rue d'Amsterdam – 75009 PARIS
  Tel. 42 85 26 30

- Syndicat Professional des Propriétaires exploitants de plates-formes d'élimination de déchets industriels (SYPRED)
  233 rue de la Croix-Nivert – 75015 PARIS
  Tel. 45 33 96 60

- Fédération Nationale des Syndicats de l'Assainissement (VANID)
  12 rue de Milan – 75009 PARIS
  Tel. 48 74 81 60

- Chambre Syndicale Nationale des fabricants de matériel de nettoiement, de collecte et de traitement des déchets (FAMAD)
  72 rue d'Amsterdam – 75009 PARIS
  Tel. 42 85 26 30

2.6 Associations

- Association Française de Normalisation
  Tour Europe – Cedex 7 – 92080 PARIS LA DEFENSE
  Tel. 47 78 13 26
- Association Générale des Hygiénistes et Techniciens Municipaux
  9 rue de Phalsbourg – 75017 PARIS
  Tel. 42 27 38 91
- Association des Ingénieurs des Villes de France (AIVF)
  91 rue Lecourbe – 75015 PARIS
  Tel. 47 83 46 39
- Progrès et Environnement
  29 rue de Berri – 75380 PARIS Cedex 08
  Tel. 42 56 82 82
- Institut International de Gestion et de Génie de l'Environnement (IIGGE)
  2 Bd de la Roche-du-Roi – BP 128 – 73101 AIX-LES-BAINS Cedex
  Tel. 79 88 20 00

3.0 Spreading of Information

3.1 Scientific and technical journals

- Compost information published by ANRED, Les Transformeurs,
  2 Square Lafayette – BP 406 – 49004 ANGERS Cedex
- Info Déchets Environnement et Technique
  21 rue d'Hauteville – 75010 PARIS
- La Récupération
  41 rue d'Auteuil – 75016 PARIS
- TSM Techniques Science Méthodes published by AGHTM –
  9 rue de Phalsbourg – 75017 PARIS

All the Industrial Waste Exchanges publish separate bulletins at varying intervals. The following are of particular interest:

- Ile de France Industrial Waste Exchange;
- APORA – Liaison (Rhône-Alpes);
- Apirec (Picardie)

3.2 Events

In addition to the Environment Shows: SITE 85 held in Paris, POLLUTEC 86, Lyon, events on solid waste may take the form of Symposiums organized by ANRED – Transformers:

- Solid Waste Management in Developing Countries (Sept. 9–10, 1987);
- Agricultural Recycling of Treatment Plant Sludges (October 27–29, 1987);

or Days (e.g. Technical Discussions of the IIGGE in Aix-les-Bains):

- Urban refuse (October 1, 1987);
- Urban Sewage Treatment Plant Sludge (December 2, 1987).

3.3 Training

A technical book by the Direction of Water Pollution and Hazards Prevention, with details of these training courses is under preparation. Orders should be addressed to the Ministry of the Environment.

3.4 Documentation centres specialized in the field of solid waste

- Centre national de Documentation sur les déchets ANRED – Les transformateurs
  2 Square Lafayette – BP 406 – 49004 ANGERS Cedex

This centre publishes a two-monthly bulletin "Solid Waste Documentation" and has organized a bibliographic data base "SCARABEE".

- Centre de Documentation sur les Déchets – Ministère de l'Environnement
  14 Bd du Général Leclerc – 92524 NEUILLY-sur-SEINE Cedex

This centre participates in the Ministry of the Environment's "Bulletin de Documentation".

4.0 Production and Composition of Solid Waste

The French Ministry of the Environment has drawn up a very accurate nomenclature of the different types of solid waste (per category and origin) that serves everywhere for the supervision of industrial waste (see Chapter 7).

The amounts of waste generated (1984) are presented more simply in Table 1.

The main trends with regard to household refuse are as follows:

Table 1 Amounts and kinds of wastes in France (1984) (million t/year)

1. Household waste:	
• Household refuse	15.0
• Outsize objects	1.5
• Old motor vehicles	1.0
• Worn tyres	0.3
Total	17.8
2. Industrial waste:	
• Inert waste (mineral tippings, quarry waste, spoil, rubble)	100.0
• Similar household refuse	32.0
• Special industrial waste possibly containing toxic matter (mercury, spent oils etc.)	16.0
• Special toxic waste	2.0
Total	150.0
3. Organic waste:	
• Of agricultural origin (cropping by-products, animal droppings), industrial (animal or vegetable products) or domestic (treatment station sludges, restaurant waste)	400.0
Total	568.0
Total wastes	735.8

- Stability of weight: small increase for coming years (about 3%);
- Increase in volume;
- Less fine and fermentable elements;
- Steady increase in the proportion of packing material;
- Increased amount of plastic, hence heat production capacity of household refuse.

Table 2 reflects the average content of household refuse in France.

However, it appears more and more necessary to possess deeper knowledge of the various sources, on a case-by-case basis and bearing local parameters in mind, in order to optimize the technical and economic choices with regard to collection (particularly selective collection) and treatment methods.

Table 2 Composition of household refuse (percentage weight %)

Paper board	20 to 35
Animal and vegetable matter	15 to 35
Glass	5 to 10
Metal	5 to 8
Plastic	3 to 6
Textiles	1 to 6
Fines (20 mm)	10 to 20
Heat producing capacity (HPC)	1200–1900 k/cal/kg

5.0 Collection of Urban Refuse

Since 1982, over 98% of the population is endowed with a refuse collecting system. The situation varies, of course, from one county to another, depending on the density of the population, the kind of housing and geographical problems (e.g. mountains).

The 2% of the population, whose rubbish is not collected are scattered over more than 5000 municipalities. Progress at the margin for these last few percent is slower and more difficult to achieve. It relies on the research for technical and economic solutions adapted to small settlements, remote from the centre of the township or village.

On the other hand, municipalities are henceforth devoting much effort to the improvement of technical and financial household refuse collection and on the development of special collections (outsize items, selective collection).

With regard to the first of these orientations, the generalization of the use of mobile dustbins and disposable bags can be seen everywhere. The installation of transit stations is developing and, lastly, new tools, such as refuse trucks with automatic grab devices, computerized circuits, etc. are commonly used.

The collection of outsize items has brought about, in particular, the development of "wasteries", centres to which the individual can voluntarily bring such waste, open to the public night and day (about 200 by the end of 1987, to satisfy 10% of the population).

The recovery and reuse of household glass have been considerably developed as part of the "packaging" contract signed between different members of the liquid food packing profession for the period 1979–1984.

The selective collection of household glass is organized by 13 902 municipalities, or intermunicipal boards, covering some 40 million inhabitants, i.e. 75% of the French population.

The breakdown of the tonnage collected by different collecting methods is:

door-to-door	62%;
voluntary delivery in small containers	26%;
voluntary delivery in large containers or on storage areas	12%.

The average collected-volume to served-inhabitant ratio is 800 g/person/month or 9.6 kg/person/year.

A new "packing" contract is currently being drafted in cooperation with the professions concerned. In accordance with the European guideline on liquid food packings of June 27, 1985, it should cover the years 1987–1990.

Table 3 Development of selective collection

	1979	1980	1981	1982	1983	1984	1985	1986
• Household glass								
Tonnage collected (10^3)	123	193	263	340	382	413	463	477
Population concerned[a] (10^6 inhabitants)	14	20	26	32	38	40		
• Paper								
Tonnage collected	—	—	—	—	150 000	150 000 to 200 000		
• PVC (bottles)								
Tonnage collected	415	1000	1130	1720	1850	1900		

[a] Population served by door-to-door glass collection or skips in street, storage areas etc.
Source of information: National Agency for Waste Disposal and Recovery Manufacturers Research and Standardization Group.

The selective collection of old papers launched in 1974 by private initiatives, has developed mainly since 1983 as part of the "old papers" contract. This arrangement set an overall target for the recycling of old papers by the end of 1986 (reuse rate of 43% instead of 37% as in 1981).

A survey carried out by COPACEL showed that in 1983, selective collection represented approximately 150 000 t. Breakdown is as follows:

	(t)
• professional recovery organizations:	30 000
• associations and schools:	9000
• municipalities:	30 000

The total amount recovered was 2 160 000 t that same year.

A new "old papers" contract is, at present, being negotiated. It will endeavour more particularly to improve the profitability of selectively collecting old papers, both for the paper trade and for the recovering establishment.

Other selective collections are regularly organized in some districts, for instance PVC or textiles. Lastly, the sorting of toxic waste, scattered among other products, at the point of output is in progress for mercury batteries, spent automobile oil or outdated medicines.

6.0 Treatment of Urban Refuse

From 30% in 1970, the rate of population disposing of treatment facilities (authorized under the legislation of July 19, 1976 on classified installations) reached 91% in 1985. (According to the 4th survey of household refuse treatment facilities completed in 1986).

Table 4 divides up the 850 licensed facilities now operating (as against 715 in 1982), according to the type of treatment applied to the household waste. Sanitary landfills under 10 t/day are not included in Table 4 and are separately reported in Table 5.

The most meaningful developments over the last six years have been:

- a slowing down in the commissioning of small-capacity incinerators, compared with the period 1975–1981;
- a sharp increase in the number of incineration plants with heat recovery equipment: almost twice as many;
- stability in the number of composting plants;
- sharp decline in refuse grinding that developed quickly between 75 and 81;
- twice the number of compacted, sanitary landfills, partly due to the use of compactors on the conventional type of sanitary landfills.

We also note the development of transit stations (nearly 150 in 1987), a natural consequence of the grouping of municipalities and the gradual abandon of the old municipal dumps in favour of the more efficient, but more remote and fewer treatment centres now available.

Lastly, for several years now, new forms of recycling household refuse have developed, as a complement to the recovery of material by selective collection (manufacture of stockable fuel, recovery of methane). This is, of course, in addition to the more conventional types of upgrading (compost for farming, recovery of incineration heat).

Table 6 shows that over one-third of the household refuse treated is effectively recycled. (One should normally add methane recovery, already in full swing on ten or more landfills, and for which a better recycling performance is under study.)

Regulatory instruments for sanitary landfills and household refuse incinerators have been recently updated and strengthened. At the same time, they have also been normalized on a nationwide scale.

A new technical order concerning sanitary landfills for urban solid waste,

Table 4 Treatment and disposal methods for urban refuse

Treatment method	No. of units	Average treated amount (t/day)	Increase of treated amount compared with 1975 (%)	Increase of treated amount compared with 1981 (%)	Population served	Fraction of population served in 1985
Incineration	218	7136	+52	−3	7 129 000	13.1
Incineration with energy recovery	66	12 296	+27	+9	12 400 700	22.8
Composting	90	4053	+24	−13	4 352 920	8.0
Grinding and WD fuels	133	7312	+108	−9	4 671 850	8.6
Landfill*	341	18 445	+110	+5	18 664 030	34.4
Miscellaneous	2	165	—	—	150 700	0.3
Total	850	46 407	+63	+2	47 369 200	87.2

* Sanitary landfills authorized under the law on classified facilities receiving over 10 t/day. Sanitary landfills for under 10 t/day should be added to the above.

Table 5 Sanitary landfills receiving under 10 t/day

	No. of units in 1985	Average treated amount (t/day)	Population served in 1985	Fraction of population served (%)
Landfills receiving under 10 t/day	700	2100	2 200 000	4

Table 6 Treatment methods for the recycling of household refuse

Methods of treatment	No. of units	Treated tonnage (t/day)	Tonnage (%)
Incineration with energy recovery	66	12 296	26.5
Composting	90	4261	9.2
• Waste derived fuel	8	318	0.7
• Pyrolysis	1	150	0.3
• Methane production	1	15	0.03
Subtotal	166	17 040	36.7
Incineration	218	7136	15.4
Grinding	125	3786	8.2
Landfill	341	18 445	39.7
Total	850	46 407	100

updating the regulations of 1973, was published in the form of a circular dated March 11, 1987. It mainly relates to:

- site layout and landscaping;
- control and management of water; management of fermentation gases;
- control of incoming waste to avoid the ingress of special industrial wastes in sanitary landfills for urban waste;
- postoperational landscaping and control.

The by-law of June 9, 1986 concerning incineration plants for household refuse introduces, among other points, more stringent specifications concerning emissions to the atmosphere. For facilities with a capacity exceeding 6 t/h or comprising a furnace with a capacity of over 3 t/h, for example, the stack gases must contain no more than:

- 50 mg/nm^3 of dusts;
- 100 mg/nm^3 of hydrochloric acid;
- 10 ppm of gaseous hydrocarbons
- 5 mg/nm^3 of total particulate heavy metals (Cu, Pb, Zn, Ni, Cr, Sn, Ag, Co, Ba);
- 0.3 mg/nm^3 of mercury and cadmium (Hg + Cd) particulate and gaseous;
- 1 mg/nm^3 of arsenic (As).

With regard to household-refuse derived compost, the approach is more of the incentive type in order to encourage high quality products on the market. In addition to the already existing and imposing standard specifications on fermentation (four days at 60°C minimum), curing, organic and nitrogen contents, we now have a label of quality awarded to composts with the characteristics reported in Table 7.

Table 7 Quality standards for compost

	Class A	Class B
(1) Granulometry of compost (square mesh screen in mm)	≤40	≤40
(2) Impure contents (% of dry solids)		
• plastic & expanded polystyrene film larger than 5 mm round mesh	≤1.6	≤2.5
• heavy particles larger than 5 mm round mesh	≤6	≤12
• total inert matter	≤20	≤35
(3) Heavy metals (in ppm of dry solids)		
• Lead	≤800	≤800
• Mercury	≤8	≤8
• Cadmium	≤8	≤8
• Nickel	≤200	≤200

7.0 Recent Installations for Reference

An accurate list of all household refuse treatment facilities is published at regular intervals by the Journal TSM (see above). The fourth and last edition dates from the end of 1986.

8.0 Industrial Waste

Industrial activity in France generates ten times more waste annually than is produced by households:

- About 100 million tonnes of inert waste coming from the extractive industries or consisting of cuttings and rubble;
- About 30 million tonnes of "inoffensive" waste that can be treated in the same way as household refuse;
- About 20 million tonnes of "special" waste containing dangerous substances at varying concentrations. It is therefore hazardous for the environment to a certain extent and special precautions must be taken with regard to its disposal. Among special wastes, a distinction is made to toxic and dangerous waste. Their disposal is subjected to special control because of the high concentration of poisonous elements. The production of such waste is estimated to be some millions of tonnes per annum. These figures originate from the inventory of industrial waste drawn up in 1979, containing the substance of three types of survey:
  - Twenty regional surveys covering all the economic activities in a given geographical area, such as a region or county, with a view to planning the treatment of rejects.
  - Twenty-two surveys per branch of activity the purpose of which was to determine the factors that generate waste in specific industries such as paper, glass and automobiles, etc.
  - Studies per product intended to evidence all the problems connected with the generation and disposal of a given residue.

The drafting of this report from scattered information that was not always easy to analyse meant, in particular, drawing norms from the regional inventories and extrapolating the results to regions where no survey had been made. It also involved finding common denominators for the meaningful results obtained through surveys per branch of activity.

The summary of these lists established in 1979 recorded 18 million tonnes of waste classified as follows:

* **Waste classified as toxic or liable to contain toxic matter**
- Salts and hydroxides
- Metalloids and compounds
- Phenol compounds
- Other organics
- Chemical waste
- Toxic waste
- Various baths

The main industrial sectors that generate toxic waste are the metal industries, textile industry, chemical and parachemical industries.

* **Waste classified as pollutant**
- Paints, inks
- Baths from textile mills and allied activities

- Solvent waste
- Miscellaneous oils
- Soluble oils
- Tar
- Other tar based and bituminous waste
- Pickling sludge
- Dairy waste
- Slaughterhouse waste

Pollutant waste consists mainly of dairy waste in the agrofood industry and oils used in metallurgy. These two kinds of residue are recycled or treated.

The industrialists who are responsible for disposing of their own waste resort to either collective or individual means. Collective facilities often consist of collective centres for specialized treatment, by incineration or detoxification (700 000 t/year installed capacity), or specialized collective dumps (the quantity of special waste received in such facilities is estimated at 500 000 t/year), or again collective recycling plant (e.g. regeneration of solvent oils, etc.) to the extent of some 250 000 t/year. There are also manufacturers that handle the disposal of their waste privately by installing plant similar to the facilities provided by collective disposal centres.

Overseas trade in toxic and dangerous waste takes place on a very small scale. Exports and imports combined represent only 5% of the national activity.

Collective means

Waste is treated in different types of plants:

- Thermal stations with a capacity of 50 000 t/year;
- Ten specialized units, scattered over French territory, treating about 200 000 t/year;
- Five cement factories licensed to incinerate special waste (notably hydrocarbons with a high heat producing capacity);
- Five centres specialized in physico-chemical treatment offering a capacity of 240 000 t/year;
- Ten units specialized in the treatment of cutting fluids and metal processing fluids, often called soluble oils and/or water–hydrocarbon mixtures, offering a total capacity of 100 000 t/year;
- Thirteen special landfills currently receiving 500 000 t/year of special waste and installed on sites where the substratum is naturally impermeable or has been artificially waterproofed.

Individual means

- The quantity of waste treated by incineration or physicochemical means is in the region of 500 000 t/year spread over and about 40 facilities on the site of the factories producing the waste.

Table 8 Recapitulatory table of special industrial waste disposal and recycling plants

	Type of unit	No. of units	Capacity (× 10^3 t/year)	Treated 1982[a] (× 10^3 t/year)	Imported (× 10^3 t/year)
Collective disposal and dumping	Incineration • In specialized collective centre	10	210	200	13
	• In thermal station	1	50	50	17
	• In cement works	5	120	15	—
	• At sea	3	—	11	—
	Sub total	—	380	276	—
	Physico-chemical treatment • In collective centre, detoxification, neutralization, sludge dewatering	5	240	130	2.2
	• Treatment of soluble oils and mixed water/hydrocarbons	10	100	90	—
	Sub total	15	100	90	—
	Sanitary landfill	13	—	500	41
Private	Internal disposal • incineration • physico-chemical	—	—	500	—
	Landfill	80	—	2000	—
	Exported (other than incineration at sea)	—	—	5	—
	Regeneration of solvents	18	70	45	—
	In collective oil centres	—	—	110	—
	Recapitulation collective and private • Incineration treatment export regeneration	—	—	1200	—
	• sanitary landfill	—	—	2500	—
	Total	—	790	3700	73.2

[a] Including imported amounts.

- As regards dumping by industrialists privately, nearly 150 deposits have been listed, receiving over 2 millions t/year.

9.0 Legislation and Regulations

The main texts concerning the recycling and disposal of waste are, for household refuse:

- Law of July 15, 1975, No. 75–633 concerning the disposal of waste and the recovery of waste materials published in the French "Journal Officiel" of July 16, 1975 and the following texts of enforcement:
  - By-law 77–151 of February 7, 1977 on the obligations of local authorities published by the French "Journal Officiel" of February 20, 1977.
  - Circular No. 77–125 of May 18, 1977 concerning the disposal of household refuse, published in the French "Journal Officiel" (J.O.) of July 9, 1977.
- Law of July 19, 1976, No. 76–663 concerning classified installations for the protection of the environment published in the French J.O. of July 20, 1976 (this law annuls and replaces the law of December 19, 1917 concerning dangerous, unhealthy or inconvenient establishments) and the texts of enforcement recently updated:
  - Decree of June 9, 1986 concerning the incineration of household refuse.
  - Circular of March 11, 1987 concerning the dumping of household refuse.

Regulations on industrial waste:

- Law of July 15, 1975, Article 8. The executive order of August 19, 1977 lays down the list of nuisance generating waste products and makes it possible to oblige firms that product, transport or discharge this type of waste to submit a regular declaration, keep a register, make out a loading statement for all transported waste specifying, in particular, the ultimate means of disposal provided.

  The by-law of January 4, 1985, concerning the control of disposal circuits with regard to nuisance-generating waste, enforces these procedures. The by-law of July 5, 1983 sets the conditions in which toxic and hazardous waste can be imported.
- Law of July 15, 1975, Article 9. Article 9 of the above law provides the possibility of fixing the conditions in which certain toxic and dangerous waste disposal activities can be pursued; it provides that the waste can only be treated in facilities for which the operator holds a permit.

  The by-law of November 21, 1979, amended March 29, 1985, concerns the recovery of spent oils, and the executive order of February

2, 1977 concerns the disposal of the waste containing PCB provides for the application of Article 9 of the law of July 15, 1975.

- Law of July 19, 1976. Different instructions giving the compulsory technical prescriptions applicable to installations receiving industrial waste (item 167) were addressed to the government officials of the various counties. They concern:
  - The dumping of industrial waste, January 22, 1980, October 16, 1984;
  - The incineration of industrial waste, March 21, 1983;
  - Public information;
  - Transportation and pretreatment of industrial waste, 30 August, 1985.

Furthermore, specific technical instructions for certain categories on the list of industrial activities may contain provisions concerning waste (surface treatments, manufacture of asbestos-based linings).

- Law of July 12, 1977. The executive order of February 13, 1985 enforcing the control of chemicals provides that, in the technical documents attached to the declaration of a chemical being offered for sale on the market, all information must be included concerning the possibilities of disposal or recovery at the various stages of its use.

To supplement the regulations published, the nomenclature of waste products published in the J.O. of May 16, 1985 was drafted in the aim of:

- providing a better definition of the different waste products;
- achieving more efficient management and closer control of what happens to the waste during the production, transportation and disposal process;
- obtaining standardized acquisition and processing of the statistical data required for research on the most suitable disposal and recycling techniques, as well as on the definition of the need for new facilities;
- establishing a common language to be used by all those who share the problems connected with waste: industrialists that produce, collect, discharge and recycle; local communities (municipalities, county and regional authorities); administrative departments and public institutions (Basin Authorities, ANRED); international associations and partners.

This nomenclature contains a double classification: all waste is described according to the association of two factors: i.e. the category it belongs to and the activity by which it is generated (origin).

(1) The category to which the waste belongs: 100 categories were determined on the basis of several criteria: physical state, chemical family, environmental hazard, treatability, recycling value.

(2) Activity by which the waste is generated (origin): the classification of waste-generating activities is derived from the group called "Level 40" of the Nomenclature of Activities and Products established by INSEE.

For waste generated by the manufacture of chemical products, extra information concerning the chemical nature of the most characteristic substance likely to make the waste toxic was incorporated in the list of generating activities.

7

Hungary

Surface: 93 625 km^2
Population: 10 657 000 inhabitants

- Number of inhabitants in Budapest: 2 071 500 (19.4%)
- Number of inhabitants in cities: 3 936 600 (37.0%)
- Number of inhabitants in communities: 4 649 300 (43.6%)

1.0 Waste Treatment and Administration

Hungary has been involved in the development of the nature and environmental protection system for a long time. The National Nature Protection Agency was established in 1969, mainly the areas declared to be protected constitute its scope of activities. The Law II of 1976 provided a new basis for the Hungarian environmental and nature protection policy. The National Environmental and Nature Protection Agency (OKTH) has been founded on the basis of the decision of the Council of Ministers and the National Environmental and Nature Protection policy was put onto a new constitutional basis. The local offices of the Agency, the county inspectorates were founded and secretaries for the County Environmental Protection agencies were appointed. The National Environmental and Nature Protection Agency has a national scope of activities. However, the guidance of higher level for the environmental protection policy and the responsibility for the actions taken are shared between government offices and national institutes.

Waste treatment is one of the most important areas of the environmental activity. Waste treatment can be defined as a unified, synchronized, vertical technological system including the collection of the wastes from the site of formation, the transportation (temporary storage, possible pretreatment), the neutralization and utilization of the wastes. The collection and processing of waste in the form of a regular and organized service meeting the requirements of public health and environmental protection are the fundamental goals of the municipal operations. These duties also constitute part of the infrastructure.

2.0 Waste Materials of Industrial Production

The first comprehensive survey on industrial waste materials was compiled in 1980. This survey was based on data supplied by the companies. Both quantitative and qualitative parameters were included in the survey. On the basis of the survey, industrial production waste materials can be estimated to be about 20 million m^3 and 25% of the waste materials can be considered dangerous. The data of the survey are summarized in Table 1, according to the industrial branches.

Table 1 Industrial wastes formed during production according to the industrial branches (1980)

Industrial branch	Amount formed (kt/year)	Percent of the total amount formed	Amount utilized (kt/year)	Percent of the amount formed
Mining	5317	24.0	48	0.9
Electric power industry	5735	25.9	580	10.1
Metallurgy	5785	26.1	3145	54.4
Machinery industry	317	1.4	168	53.0
Chemical industry	390	1.8	192	49.2
Light industry	199	0.9	64	32.2
Food industry	1900	8.6	500	26.3
Building material industry	2503	11.3	726	29.0
Total	22 146	100.0	5423	24.5

Table 2 Nonutilizable industrial wastes (1983)

Designation	Amount (t/year)
Galvanic sludge	32 000
Catalysts	260
Metal-containing wastes	3712
Cellulose-containing wastes	6400
Fat-containing wastes	2500
Wastes, burnable without pretreatment	11 340
Wastes, burnable with pretreatment	25 700
Mixed, solid waste	17 000

The Industrial Ministry made a survey in 1983 on waste materials of industrial production for which wastes the utilization, sale and disposal remained unsolved.

The National Environmental and Nature Protection Agency keeps a register on the dangerous waste materials on the basis of data supplied by the companies. This can be considered approximately complete for the industry. The majority of dangerous waste materials originates from the aluminium industry, chemical industry, food processing industry, metallurgical and machinery industry.

The distribution of dangerous wastes according to the industrial branches and their percent contribution are given in Table 3. The distribution according to the origin is presented in Table 4.

Table 3 Distribution of dangerous wastes according to the industrial branches (1983)

Industrial branch	Amount (kt/year)	Percent of the total amount
Mining	193.0	5.48
Electric power industry	1225.5	34.80
Metallurgy	1452.8	41.25
Machinery industry	109.1	3.10
Building material industry	8.6	0.24
Chemical industry	136.0	3.86
Light industry	143.4	4.07
Food industry	253.0	7.18
Others	0.6	0.02
Total	3522.0	100.00

Table 4 Dangerous industrial wastes according to origin (1983)

Origin	Amount (kt/year)	Percent of the total amount
Wastes of plant and animal origin	311.0	8.83
Wastes of mineral origin	2340.0	66.44
Wastes formed during chemical conversion	591.0	16.78
Other wastes	280.0	7.95
Total	3522.0	100.00

2.1 Collection, treatment and disposal of industrial wastes

Broadly speaking, the collection of dangerous waste materials is solved and organized everywhere. However, the storage, neutralization and possible reuse are not worked out completely. About 30% of the industrial wastes listed in the dangerous waste material register are taken for neutralization and processing.

The regulation for the shipping of goods of the Hungarian State Railways, the supplements "A" and "B" of the European Agreement for the Transportation of Dangerous Goods on International Public Highways and the regulations and ordinances prepared on the basis of the above are the guiding rules for the transportation of dangerous industrial wastes. It is a principle that license is required for the transportation. Closed containers having a volume of 4 and 5 m^3 and in certain cases, barrels are used for the transportation of solid dangerous wastes.

A major share of the dangerous wastes is stored at disposal sites of the habitational refuse or at the disposal sites of the nondangerous wastes at the companies. In certain cases, the wastes are burnt in boilers or in small amounts in open areas. 20% of the waste (300 000 t/year) which deserves increased attention is neutralized by the proper method at the companies. The burning of the wastes formed is carried out in a number of incinerators, however, only three up-to-date units are capable of handling dangerous wastes with changing composition.

These are as follows:

Leninváros, TIFO:	Ghelfi type rotary kiln with afterburning furnace and flue gas purifier. Nominal output: 2 t/h (17 GJ/h)
Százhalombatta, DKV:	Purator type rotary kiln with afterburning furnace and flue gas purifier. Nominal output: 5 t/h (37.5 GJ/h)
Balatonfüzfő, Nitrokémia:	Purator type rotary kiln with afterburning furnace and flue gas purifier. Nominal output: 1.45 t/h (13.7 GJ/h)

The total output of the three incinerators is under 10–12 t/h. The incinerators are in operation for about 7 months per year.

The Jászberény Chipping Machine Co. manufactured some Purator mouth-type incinerators for hospitals (Szolnok, Szigetvár, Budapest). These are capable only of handling infectious wastes of hospitals with a nominal output of 100–400 kg/h.

On the basis of the survey of the industrial ministry incinerators incorporated directly into the technological system can be found at 22 companies and incinerators (Ghelfi, MAK-I, PURATOR, SUMITO, PS-SUS-II types) are under preparation and construction at five companies. The equipments cannot be leased because of their high capacity utilization.

Some big enterprises have set up disposal sites within their own scopes for the organized disposal of inorganic, dangerous wastes. The following companies can be mentioned:

- Industrial Works of County Hajdu; Hadháztéglás;
- Hungarian Optical Works;
- Tungsram RT.

The following regional projects are under planning and construction:

- The development of a national network for the treatment of dangerous wastes (three incinerator plants, six disposal sites and county transit storage facilities;
- The development of neutralization facilities at the companies;
- The introduction of technologies with less amount of waste. These steps could result in a satisfactory solution.

However, the costs of investments exceed the limits of the five year planning period. The project to be realized is the following according to the extended range planning of the national economy:

- One incinerator plant (Dorog) and one disposal site (Aszód) are to be put into operation;
- One incinerator plant (Rudabánya), two disposal sites (Garé-Bosta, Vál-Vértesacsa) and county transit storage facilities have to be initiated between 1989–1992 as well as the development of technologies with less amount of wastes and the improvement of the waste utilization.

2.2 Agricultural wastes

The survey of the wastes which are formed during agricultural production and the processing of the agricultural products was done by the Ministry of Agriculture and Food Administration in 1985. Twenty eight wastes of plant and animal origin were included in the survey. The survey was rendered more difficult by the fact that unappreciable questionnaires were sent back from 10% of the agricultural plants and from 70% of the food processing plants.

The results of the survey are included in Table 5. However, it should be noted that only the relative values can be considered to be competent. There is no available data on the amount of industrial wastes formed during agricultural production. A significant part of the wastes can be considered to be dangerous (residues of plant protectives, packing materials of chemicals). The amount of solid waste formed in agriculture is estimated to be about 30–35 million t/year.

Table 5 Amount and utilization of the more important solid agricultural and food industrial wastes and by-products

Designation of the waste	Amount formed (kt/year)	Utilizable amount (kt/year)	Utilizable amount related to the amount formed, percent
Grain straw	4357	4246	97.45
Corn stalk	4655	4605	98.93
Corn cob	101	99	98.02
Leafy turnip	892	889	99.66
Fruit, grape trimmings	146	138	94.52
Meat waste and haslet	141	139	98.58
Animal carrion	75	73	97.33
Horn hoof	0.3	0.3	100.00
Vegetable waste	83	75	90.36
Sawdust	36	33	91.67
Manure	12 505	12 373	98.94

3.0 Habitational Solid Refuse

Habitational solid refuse can be defined as a mixture of organic and inorganic materials having different sizes and compositions, formed in residential buildings of settlements, public institutions, areas of public traffic and public parks. The garbage removed from producing plants, which has a similar composition to the household and institutional refuse and which is collected together with habitational refuse as well as the waste materials formed during construction are included in this class too.

More than 14 million m^3 of habitational solid refuse should be taken care of in the country regarding collection, transportation and neutralization. The quantity and composition of the habitational solid refuse are changing according to the living standards, way of life and consumption practice. Only estimations are available for the total amount of the habitational solid refuse formed in the country, however, data are available for the amounts of the wastes collected, transported and neutralized between the organized structure.

The organized waste collection has been spread over for about nearly 800 settlements. The planned development of waste collection, transportation and utilization and the selection of methods according to environmental, ecological and economical standpoints should be based on reliable basic planning data regarding the quantity and quality of the waste.

The Environmental Protection Department of the Budapest Public Territory Maintaining Company continuously monitors the change of the amount, the morphological, physical, chemical, biological and ecological characteristics of the refuse from residential buildings, the wastes from commercial activities, public service and the wastes of no technological origin from producing plants.

On the basis of the results, the following conclusions can be made: the relative amounts of papers, plastics, and metals shows an increasing tendency regarding the major constituents; the relative amount of the organic components decreased; the heat value of the refuse meets the requirements of the economical combustibility; the combustible material content and the proportion of the constituents which could be used as potential secondary raw materials showed an increasing tendency (Table 6).

Table 6 Composition, density and heat value of the habitational solid refuse formed in Budapest (1970–1986)

Designation of the waste material component (%)	1970	1975	1980	1985	1986
Paper	18.1	25.2	23.2	16.25	19.98
Textiles	2.5	2.4	3.9	4.08	5.31
Plastics	2.7	4.2	4.5	3.08	5.70
Glass	2.9	3.9	4.0	5.61	6.09
Metals	1.5	2.6	3.5	4.24	4.43
Organic waste	27.3	29.2	33.9	39.00	34.73
Inorganic waste	45.0	32.5	27.0	27.74	23.76
Potential secondary raw materials	27.7	35.4	39.0	33.90	41.50
Density (t/m^3)	0.3–0.35	0.23–0.30	0.17–0.25	0.15–0.21	0.16–0.21
Heat value (kJ/kg)	5020–6280	6700–7750	7570–7960	6890–8860	6990–7560

3.1 Collection and transportation of habitational solid refuse

The collection of the refuse is carried out in vessels with volumes of 110 l, 240 l, 1.1 m^3 and 5 m^3. The habitational refuse is transported in Budapest by modern equipment having compacting presses. In the country, the semidustless collection is widely used. The collection is carried out in an organized system which means that every collecting machine works according to a schedule and collects the rubbish from a given area. The frequency depends on the density of the population and the storage possibilities and in the area of Budapest is as follows:

Five times per week:	15%
Three times per week:	33%
Twice per week:	31%
Once per week:	21%

3.2 Methods of neutralization and utilization of habitational solid refuse

Landfilling and dumping

At present, the decisive method of neutralization of habitational solid refuse is organized or unorganized dumping. The refuse is transported by the collecting-transporting equipment to the designated areas (dumping sites, disposal sites), where the refuse is used for filling of wells, mining pits or is deposited in prism-shaped formations.

There are 275 disposal sites in the scope of the organizations involved with the collection and transportation of the refuse. However, the number of the disposal sites which belong to other organizations and the number of the illegal dumping sites are probably much higher. However, data are not available regarding this matter.

The possibilities of utilizing solid habitational refuse neutralized by dumping are restricted. Experiments are carried out at different places for the utilization of the biogases accumulated at the disposal sites. More biogas plants are now in operation, where the energy obtained is used for local purposes. The low investment cost and the prompt refund could support its wide-range spreading.

Incineration

The neutralization of a significant part of the solid habitational refuse by burning is carried out only in Budapest. The incinerator started its operation in 1982. Its planning capacity is 1200 t/day.

The dust emission of the stack gases is under the emission limit value. The refuse is burnt in incinerators at temperatures of 850–1100°C. The slag produced at the high temperature can be considered as sterile; the volume of the slag is about one-tenth of the original volume of the refuse. The disposal and the utilization of the slag at road construction is under development.

The steam generated during the incineration produces electric energy and maintains the self-consumption of the incinerator. The incinerator supplies energy to the national energy network and it supplies steam to the surrounding housing estates and industrial plants.

The iron scrap is separated by an electric magnet from the slag formed during the incineration and is taken for utilization in metallurgy.

Composting

Compost-making industrial plant can be found in Keszthely, based on the solid habitational refuse of the neighbourhood and on the drained sewage sludge of the sewage treatment plant.

There were problems during the operation of the compost-making plant since the amount of refuse in the collection area did not meet expectations. The moisture content of the refuse was much higher than calculated and the sale of the final product was accompanied by problems.

The most effective method of utilization of the solid habitational refuse is to impede the valuable constituents to get into the process of the collection, shipping and disposal of the refuse. One method is to buy up the different waste materials (paper, textile, metal, glass, etc.). The population is not encouraged in the present form to utilize the useful wastes in the household refuse. The other possibility is selective waste-material collection according to the types of the materials. However, the experiences of this were extraordinarily unfavourable in Budapest.

Table 7 Treatment and disposal methods for habitational solid refuse (1000 m^3/year)

Landfilling	13 069.3
Incineration	1615.0
Composting	74.1
Others	377.2
Total	15 135.6

4.0 Utilization of the Waste and Secondary Raw Materials

4.1 Material cost saving from the utilization of wastes

The value of wastes recycled back into production or utilized within the companies was 12.1 thousand million Ft in 1984. The value of the producing waste accounted decreased the material cost more than 1%. The determinate proportion of the waste is formed and utilized in the industry.

The full scale of waste material utilization includes not only producing waste materials but the wastes and by-products formed but not accounted in other areas of the economy and the savings from the use of these materials as substitutes.

4.2 The change of the waste material collection

The collection has a significant role in the waste material utilization beside the direct utilization and the increased trade between the companies. The major part of the aluminium and coloured metal wastes, two-thirds of the rubber trimmings and waste paper, half of the iron and steel scraps and a quarter of the plastic waste are utilized through the collecting plants. The collection has an insignificant role in the case of wood, leather and glass wastes.

4.3 Utilization of certain waste materials

Iron and steel scraps

Approximately 2.5 million tonnes of iron and steel scraps were used for the metallurgical products. 529 kg of waste materials were used for 1 tonne of raw steel and 869 kg of waste materials were used for 1 tonne of cast iron and cast steel. The increase is more than 5% relating to 1980.

Coloured metal wastes

The amount of coloured metal wastes collected has increased significantly (28–40%) on the basis of the metal content relating to 1980. The ratio of coloured metal wastes collected to the raw materials improved because of an increase in the utilization.

Rubber trimmings

The rubber trimming is made up of more than two million tyres. The renewal of tyres showed an increase in 1981 and 615 000 tyres were renewed in 11 plants in 1984. This value is more than 2.5 times the value for 1981.

Waste papers

The raw material supply of the paper industry has a strong impact on the natural environment; and the energy supply requirement of the paper industry is high too. Waste papers have been used for a long time to replace the primary raw materials.

The amount of waste used for paper and paperboard production increased by about 65 000 tonnes (30%) during a four year period and its ratio within the total fibre material attained 47.6%.

Export of waste materials

One form of utilization of wastes is the export. However, this has a supplemental role beside the national processing. The dynamics of the export of certain waste material groups represents the supplemental feature and contingency of the waste material export.

5.0 Addresses

- Ministry of Housing, Public Construction and City Planning (EVM)
  1054 Budapest, Belloiannisz u. 24.
  Tel. 532 200, 123 420

The housing and communal department of the Ministry of Housing, public construction and city planning deals with the guidance, technical supervision and technical consulting of the communal companies.

- National Environmental and Nature Protection Agency (OKTH)
  1051 Budapest, Arany János u. 25.
  Tel. 327 739

The agency organizes and follows the executions of the decisions and works out proposals for the environmental goals of the people's economic plan in cooperation with the parties concerned. The agency takes part in the organization of scientific research for the environmental protection and assures the development of the conditions.

- Budapest Public Territory Maintaining Co. (FKFV)
  1081 Budapest, Kállai Éva u. 7.
  Tel. 336 930
  Telex: 225569

Scope of activity:

- Collection, transportation, neutralization of the solid communal refuse
- Neutralization
  - organized dumping
  - incineration
- Analysis of the solid communal refuse

- Public Sanitary and Settlement Cleansing Association
  1068 Budapest, Rippl Rónai u. 34–36.
  Tel. 318 921

Coordination of the provincial settlement cleaning companies and consultation for solving the different environmental problems.

- MÉH Raw Material Utilization Co.
  1061 Budapest, Népköztársaság u. 45.
  Tel. 229 249

  The scope of the company is as follows: selective collection of the waste according to specifications, processing and utilization of the waste.

- Land Measuring and Ground Inspecting Co. (FTV)
  Department of Waste Technology
  1116 Budapest, Építész u. 40–44.
  Tel. 453 520

  The main activity during the last two years was the development of neutralization and utilization technologies of the waste materials.

  The company deals with the biotechnological methods of the decomposition of wastes on the basis of their latest experimental results.

- Nature Protection Institute
  1113 Budapest, Aga u. 4.
  Tel. 451 350
  Telex: 227465

  The main activities can be summarized as follows:
  - Treatment of dangerous wastes;
  - Evaluation of the qualitative and quantitative features of dangerous wastes;
  - Tests for the qualification of wastes;
  - Technological investigation of the neutralization, disposal and utilization of dangerous wastes;
  - Study of the effects of technologies and plants on the environmental structure and development of proposals for the solution;
  - Preparation of the environmental protection projects for city development.

  The operational area of the institute is the whole country. The organization consists of the Budapest central divisions and seven provincial, regional stations.

- "NATURA" planning, development and service cooperative society for environmental protection
  1364 Budapest P.O. Box 8
  Tel. 490 726 Ext. 25
  296 246 Ext. 25
  Telex: 225007

Scope of activities:
- Zinc removal from metallurgical dusts;
- Pyrolysis of scrap leather containing chrome;
- Utilization of the backing sand in the building material industry;
- Design of transit waste depots with modern technology meeting the requirements of the strict Hungarian standards.

- PROMETHEUS Firing Technical Co.
  1108 Budapest, Gyömrői u. 140.
  Tel. 272 018

Development of boilers which are suitable for burning different industrial wastes in an environmental protective way on the basis of own design and in co-operation with other companies.

- TÜKI-Firing Technical Research and Development Co.
  3515 Miskolc-Egyetemváros
  Tel. 46 69 822
  Telex: 62351

The company has a well-founded technical experience obtained in different national research projects and has a good supply of technical equipments.

The company is ready for delivery and commissioning of equipments for the neutralization of solid, liquid or gaseous wastes where incineration or heat treatment is the most expedient neutralization method.

- Institute of Management of Water-Supplies
  1054 Budapest, Alkotmány u. 29.
  1369 Budapest P.O. Box 352
  Tel. 329 940
  Telex: 224150

The institute has been involved in preparing plans for complex environmental protection including disposal problems of waste materials on the basis of ample international experience for about ten years.

- AGROÉP Subsidiary Co., Nyirtelek
  4461 Nyirtelek, Petőfi S. u. 3.
  Tel. 42 15 671
  Telex: 73617

The company has been manufacturing solid, liquid, dangerous, useful and waste material collecting and storage vessels and equipments of DEFT type. These are closed, cylindrical equipments for medium pressure range (1–4 bar) with a volume between 1–10 m^3.

6.0 Scientific and Technical Journals

Information Booklet on Urban Economy
Published by: Information Centre, EVM
50 pages; 5–10% of the booklet deals with solid waste materials.
Published quarterly, since 1967

Environmental Control
Published by: National Technical Information Centre and Library
60 pages;
Published monthly, since 1971

Utilization of wastes and secondary raw materials
Published by: National Technical Information Centre and Library in cooperation with the MEH Co., Metalloglubus and KOKÖV.
60 pages;
Published monthly, since July 1977

Papers on the Environmental Protection from the International Press
Published by: Hungarian News Service
70 pages;
Published monthly, since 1987

7.0 Institutes of Higher Education

Chemical Engineering School of Budapest Technical University. Two year postgraduate education is organized. The presentation of waste treatment and neutralization takes one semester. The technical–technological instruction takes two semesters.

Special courses are devoted to the treatment of communal and dangerous wastes in the Postgraduate School of the Budapest Technical University. The courses are scheduled for one semester.

Heavy Industrial Technical University of Miskolc. Postgraduate education is organized in the School of Metallurgical Engineering which takes two years. The topics are the neutralization and utilization of industrial wastes.

Veszprém University of Chemical Engineering (Veszprém). The treatment of the wastes is the main topic in postgraduate courses.

Ybl Miklós Technical College (Budapest). The educational material on the treatment and utilization of wastes is presented within the basic subjects at the Faculty of Urban Economy for the regular students. The waste material treatment is presented in a facultative class entitled "Preliminary Engineering Design and Environmental Control" at the Faculty of Civil Engineering for the regular students. Two year factory engineering postgraduate education will be initiated from the spring of 1988 under the title of "Techniques and Technology of Constructed Environment". The treatment and utilization of waste materials will be presented.

Gödöllő University for Agricultural Sciences (Gödöllő). Postgraduate courses within the environmental protection education. The soil protection and environmental protection subjects focus on the treatment of agricultural and communal wastes.

Special courses are organized and proceedings are published by the National Environmental and Nature Protection Agency to promote comprehensive understanding of the subject.

8.0 Legislation

The Law II of 1976 provides rules for the protection of the human environment. The law declares the goal as the protection of the health of the people and the improvement of the living conditions of the present and the following generations. The elements of the environment (earth, water, air, habitational environment) have received protection.

The more important ordinances regarding dangerous and habitational wastes are as follows:

- The ordinance no. 56/1981 /XI.18/ of the council of ministers on inspecting the formation of the dangerous wastes and on the activity for their neutralization.

The ordinance involves the inspection of the formation, temporary storage, transportation, neutralization of dangerous wastes, the prevention of the environmental impairing effect of the wastes and the fines regarding the above. The list of dangerous wastes and the register to be kept in file at the producing plant are given in the supplements of the ordinance.

- Publication no. 9002/1983 /TK.12/ OKTH of the National Environmental and Nature Protection Agency deals with tests on the qualification of wastes.

The obligatory use of the technical specifications for the content, methods and process of the tests for the qualification stipulated in the ordinance no. 56/1981 /XI.18/ MT. is specified.

The guide no. 8005/1984 /Eü.K.3/ EüM. of the Ministry of Health deals with the sanitary institutions which are authorized to carry out the tests for the qualification of the dangerous wastes.

The sanitary institutions are listed which are allowed to carry out basic tests for the classification of dangerous wastes and necessary sampling for a nominal fee according to the ordinance no. 56/1981 and the publication no. 9002/1183.

The directive no. 33/1984 /BK.10/ BM of the Minister of the Interior focuses on the prevention of accidents and the way of the rescue operations in case of an incident with special danger during the transportation on public roads and waterways.

The order no. 8/1983 /EüK.10/ of the Minister of Health focuses on the registration and neutralization of the dangerous wastes formed in public health institutions.

The supplement of the regulation no. 56/1981 MT gives instructions that the handling of unidentifiable wastes should be carried out by the national public health institutions, the county and metropolitan public health and epidemiological service stations. The supplement shows the dangerous waste registration sheet for public health institutions.

The regulation no. 2/1984 /I.17/ EVM of the Minister of Housing, public construction and city planning summarizes the sectoral activity regarding the inspection of the formation of dangerous wastes and neutralizaton of wastes according to the ordinance no. 56/1981 MT.

The scope of the regulation includes the collection and neutralization of dangerous wastes formed during the activities in the building industry, building material industry, constructional joinery and during the service in housing, city and municipal administration.

The order no. 10/1983 /HK.9/ of the Minister of National Defense deals with the duties regarding the formation, monitoring and neutralization of dangerous wastes.

The regulation defines the duties of the Environmental Protection Council of the Hungarian People's Army, the Head of the Ministry of National Defense, chiefs of branch of service, chiefs of service and military organizations.

The regulation no. 6/1983 /VIII.25/ Ip.M. of the Minister of Industry deals with inspecting the formation of dangerous waste materials and the actions to be taken during the neutralization of the wastes according to the execution of the ordinance no. 56/1981. The supplement of the regulation summarizes the list of the organizations under the administration of the Ministry of Industry for the inspections of dangerous wastes.

The enactment no. 5/1984 /II.7/ KM of the Minister of Transport deals with the execution of the order no. 56/1981 MT. The scope of the enactment involves the controlling of dangerous wastes formed in the transportation branch of the people's economy and deals with the neutralization of the wastes. The transportation of dangerous goods is not included in the scope of the order.

The ordinance no. 6/1983 /III.30/ MEM of the Minister of the Agriculture and food administration deals with the realization of the ordinance no. 56/1981 MT. The ordinance orders the neutralization of the plant protectives, the animal carrions, and wastes of animal origin.

The joint decree no. 1/1986 /II.21/ EVM-EüM of the Minister of Housing, public construction and city planning and the Minister of Health focuses on the activity of the public sanitation and on the habitational solid refuse. The cleanliness of the real estates, public areas, the supply of public sanitation, the neutralization and utilization of the habitational solid refuse, the orders relating to the public sanitation and the inspection of the observance of the regulations come under the ruling.

8

Italy

Surface: 301 260 km^2
Population: 56 244 000 inhabitants
Municipalities: 8081 of which:

- 4 with more than 1 million inhabitants
- 44 with 100 000–1 000 000 inhabitants
- 2059 with 10 000–100 000 inhabitants
- 5974 with less than 10 000 inhabitants

1.0 Waste Policy

Only in relatively recent years, Italy has got thorough regulations concerning the different field of environmental protection; in 1986 a Ministry of Environment (Ministero dell'Ambiente) was instituted, with the task of addressing and coordinating activities mainly committed to regional governments or other local authorities.

In recent years, the construction of new incinerating and recycling plants has marked a certain delay, also because of the growing importance of the debatement on the risks of atmospheric pollution and of the difficulties often arising from the marketing of the recovered products. A great impulse has thus been given to the realization of new sanitary landfills, on the basis of the very strict regulations adopted, at first, by the regional authorities and then included in the national regulations. At the same time, separate collection, especially of glass and paper, and more recently, on experimental basis, of tins, plastic bags and of some toxic products contained in municipal solid wastes (batteries, containers for pesticides) has been spurred.

In the last years (dating back to 1976) only 40 incinerating or recycling plants have been constructed, for a theoretical capacity of 5800 t/day on the whole, with an investment of 262 million of Italian Lire (1986 actualized value).* A similar capital has been invested for landfills (292 million of Italian Lire) for a capacity of 11 450 t/day.

About 60% of the Italian population appears to be still not provided with adequate systems for the final disposal of wastes; to satisfy the needs for waste disposal, in the next ten years an investment of some 2000 million of

*At the end of 1986, one dollar was quoted at about 1300 Italian Lire.

Italian Lire (actual values) will be required, on the basis of a policy encouraging anyway the recovery of energy and materials, overcoming the present landfill oriented strategy.

The disposal of industrial—and especially of hazardous—wastes is carried out, in most cases, by private operators: the big companies are generally provided with independent disposal structures, while the medium and small factories refer to external, private operators. In the last cases, the public control—defined by the recent regulations—has not always shown a satisfying level of efficiency.

At present, no official information is available on the effected and foreseen investments in the field of waste disposal. The Italian federation of the public services of local institutions (see Section 2.3) has recently promoted a study on this subject, from which the data reported below have been extracted, and which can be consulted for further detail:

- Publitecnica (1986): La normativa per la tutela dell'ambiente: disposizioni comunitarie, nazionali e locali, stato di attuazione e prospettive di evoluzione. Contributo per un codice dell'ecologia (Regulations for the environmental protection: communitary, national and local provisions, state of application and evolution perspectives. A contribution for a code of ecology). Published by Ente Autonomo Fiere di Padova.

2.0 Reference Addresses

2.1 ISWA National member

International Solid Wastes and Public Cleansing Association (ISWA) Sezione Italiana, Viale Isonzo 14/1, 20135 Milano. Tel. 2 5458566

- National representative and Executive Committee member: Prof. Eugenio de Fraja Frangipane
- Scientific and Technical Committee member: Prof. Luca Bonomo
  Both at the following address: Istituto di Ingegneria Sanitaria, Politecnico di Milano (Institute of Sanitary Engineering of Technical University of Milan), Via F.lli Gorlini 1, 20151 Milano. Tel. 2 4520543–4520354; Telex 333467 POLIMI I

2.2 Government offices

- Ministero dell'Ambiente (Ministry of the Environment)
  Servizio per la Prevenzione degli Inquinamenti ed il Risanamento Ambientale (Department for Pollution Prevention and Environmental Recovery). Director Prof. Vittorio Silano

Attn. dr. Morelli, Tel. 6 6785835, and dr. Incocciati, Tel. 6 6786431
Piazza Venezia 11
00187 ROMA

- Istituto Superiore di Sanità del Ministero della Sanità (Health Institute of the Ministry of Health)
  Reparto Igiene del Suolo (Department of Soil Hygiene)
  Attn. ing. Egidio Zavattiero, Tel. 6 4990
  Viale Regina Elena 299
  00161 ROMA

- Ente Nazionale per le Energie Alternative, ENEA (National Institute for Alternative Energies)
  Dipartimento Protezione Ambiente e Salute (Department of Environment and Health Protection)
  Piazza Morandi 2
  20121 MILANO

The central Government has just generally oriented coordination tasks; the planning and practical realization of waste collection and disposal systems depend on the 20 local governments (regions) in the national territory.

2.3 Technical and professional associations

- Associazione Nazionale di Ingegneria Sanitaria ed Ambientale (National Association of Sanitary and Environmental Engineering)
  Attn. ing. Appio Claudio Di Pinto, Tel. 6 4744397–486415
  Piazza Sallustio 24
  00187 ROMA

  It is essentially a consultant and technician association, whose members operate in the realm of pollution control. A special committee is being constituted for solid and hazardous wastes.

- Unione Imprese Difesa Ambiente (Environmental Protection Companies Association)
  Attn. dr. Carlo Mozzati, Tel. 2 809006
  Piazza Diaz 2
  20123 Milano

  It is an association of engineering, planning and management companies working in the field of environmental protection. A special section for solid wastes has been recently constituted. The association is operating in the frame of the Italian National Mechanical and Engineering Industry Association (ANIMA).

- Confederazione Italiana dei Servizi Pubblici degli Enti Locali (Italian Federation of the Public Services of Local Institutions), Tel. 6 314444
  Piazza Cola di Rienzo 80
  00192 ROMA

The Federation groups the local institutions dealing with the management of public services, including many of the public cleansing and solid waste disposal services.

- Assorecuperi (Recovery Association), Tel. 2 77501
  Corso Venezia 47/49
  20121 MILANO

The association gathers private operators working for the collection, manufacturing and trade of recovered materials.

2.4 Main research institutions

- Consiglio Nazionale delle Ricerche (National Council of Research)
  Progetto Finalizzato Energetica (2)—Sottoprogetto Biomasse ed Agricoltura (Oriented Project for Energy (2)—Subproject Biomass and Agriculture), Tel. 2 783051
  Piazza R. Morandi 2
  20121 MILANO

It is an agency, made up by National Council of Research jointly with National Institute for Alternative Energies (see point 2.2). The agency finances and coordinates research carried out by public institutions, private companies and university institutes in the realm of energy recovery from agricultural, industrial and municipal wastes.

- Istituto di Ricerca Sulle Acque (Institute for Water Research), Tel. 6 841451
  Via Reno 1
  00198 ROMA

It deals mainly with waste water as well as with sludge treatment and sludge disposal techniques; research on solid waste is limited to sludge treatment in combination with urban wastes.

- Istituto di Ingegneria Sanitaria del Politecnico di Milano (Institute of Sanitary Engineering of Technical University of Milan), Tel. 2 4520543–4520354
  Via Fratelli Gorlini 1
  20151 Milano

- Dipartimento di Ingegneria Idraulica ed Ambientale dell'Università di Pavia (Department of Hydraulics and Environmental Engineering), Tel. 382 391300
  Polo Universitario del Cravino
  Via Abbiategrasso 209
  27100 PAVIA

- Dipartimento di Idraulica, Trasporti e Strade dell'Università di Roma "La Sapienza" (Department of Hydraulics, Transports and Roads), Tel. 6 4741967
  Via Eudossiana 18
  00184 ROMA

3.0 Spreading of Information

3.1 Scientific and technical journals

- Ingegneria Sanitaria (Sanitary Engineering), edited by National Association of Sanitary and Environmental Engineering (see Section 2.3), c/o Dipartimento di Idraulica, Trasporti e Strade dell'Università di Roma "La Sapienza" (see Section 2.4); six issues per year for about 500 pages, 15% of which deals with solid wastes. It was first published in 1952;

- Rifiuti Solidi (Solid Wastes)
  CIPA srl
  Viale Isonzo 14/1
  20135 MILANO

Six issues per year, for about 600 pages, entirely dealing with solid wastes. First published in 1987;

- Ingegneria Ambientale (Environmental Engineering), edited by CIPA srl (see above)

Nine issues per year, for about 720 pages, 30% of which dealing with solid wastes. First published in 1972;

- Inquinamento (Pollution)
  Via Nuovo Rivoltana 95
  20090 LIMITO (Milano)

Ten issues per year for about 1000 pages, 15% of which deals with solid wastes. First published in 1958;

- TSP—Tecnologie dei Servizi Pubblici (Public Service Technologies). Journal of Confederazione Italiana Servizi Pubblici degli Enti Locali (see Section 2.3)
  Gruppo Editoriale Stammer
  Centro Commerciale San Felice
  20090 SEGRATE (Milano)

  Six issues per year, for about 550 pages, 15% of which deals with solid wastes. First issue appeared in 1981;

- L'informatore del recupero (Information on Recovery). Journal of Assorecupero (see Section 2.3)

  12 issues per year for about 480 pages, all dealing with the technical and commercial aspects of recovery. First published in 1979.

Indications are given only for the most important journals or for those especially dealing with solid waste problems. All the above mentioned journals are Italian written and, except for the last one, they all provide an English and a French summary for each paper.

3.2 Exhibitions

No specific exhibition exists exclusively dealing with solid wastes. The subject is normally included in more general events concerning the various aspects of environmental protection. The most important among them, taking place regularly, are:

- Tau Expo (Tecnologie per la salvaguardia dell'Ambiente e la protezione dell'uomo), organized by:
  Tau Expo—Etas Prom srl
  Via Mecenate 91
  20138 MILANO
  Tel. 2 5075. Telex: 331342 ETASPE I

The exhibition takes place at the Fiera di Milano and, starting from 1987, has replaced another exhibition, the Disinquinamento–Antinquinamento, the eighth edition of which was held in 1985. In March 1987 more than 500 exhibitors, 130 of them foreigners, participated in the Tau Expo, on an area of 35 000 m^2.

- SEP Pollution: Città e Ambiente (Town and Environment), organized by:
  Ente Fiere di Padova
  Via N. Tommaseo 59
  35131 PADOVA
  Tel. 49 840111. Telex: 430051 FIERPD I

The exhibition takes place at the Fiera di Padova and its eleventh edition was held in 1986, with more than 400 exhibitors on an area of about 60 000 m^2.

3.3 Conferences

Conferences take place regularly on the occasion of the above-mentioned exhibitions (see Section 3.2). These conferences always include one session (two or three half-days) on solid wastes and are Italian spoken. In some of the most recent editions, for some sessions, a simultaneous translation into English was provided.

Detailed notice about conferences and symposium, taking place at no regular interval, is reported in the journals "Ingegneria Ambientale" and "Ingegneria Sanitaria" (see Section 3.1).

3.4 University courses

Solid wastes are generally included in the programme of the course of Sanitary Engineering, existing in the Faculty of Civil Engineering of the main Italian universities. The only university course entirely dealing with solid and toxic wastes, called Complements of Sanitary Engineering, is carried out on a yearly basis, since 1984, at the Institute of Sanitary Engineering of the Technical University of Milan (see Section 2.4).

4.0 Production and Composition of Solid Wastes

According to the provision of the law (1982 and 1984, see Section 8), wastes are classified as follows:

- Municipal wastes, originating from civil buildings and settlements, bulky and street cleansing wastes included. Also included in this class are wastes of different origin, if their properties are such that the common facilities for municipal wastes can provide their adequate disposal;
- Special wastes, including the residues from industrial, agricultural and commercial activities and from hospitals and health-care establishments, not included in the municipal waste class; those deriving from demolitions, constructions and excavations; obsolete machinery and instruments and vehicles out of use; the residues from waste treatment and sewage sludges;

- Hazardous wastes, containing toxic substances in such concentrations to be hazardous for the environmental health; the list of toxic substances and the relevant limit concentrations are reported in the tables annexed to the above-mentioned provision of the law. The wastes originating from a series of industrial activities and services are also included in hazardous wastes, unless the producer can prove they are harmless.

Reliable information on waste production and composition, collected by public or publicly controlled organizations, are only available for municipal solid wastes. For the other kinds of wastes, the available data were collected before the regulations now in force were issued, and thus do not reflect the above-mentioned classification. Information on this subject will be given in Section 7.

For municipal solid wastes (MSW), the results of a comprehensive investigation carried out per sample all over the national territory were published in 1980 by the National Council of Research (CNR). The entire data can be found in:

- Consiglio Nazionale delle Ricerche—Progetto Finalizzato Energetica:

Table 1 Patterns of municipal waste quantities and properties in the various geographic areas of the Italian territory, after the data published by CNR in 1980

Geographical areas	North-West	North-East	Centre	South	Islands	National value
Inhabitants (millions)	15.4	10.3	10.7	13.4	6.4	56.2
Main cities	Milan	Venice	Rome	Naples	Palermo	—
Waste generation (kg/cap/day)	0.64	0.69	0.69	0.73	0.64	0.68
Chemical and physical properties (%)						
• water content	40.3	41.3	42.9	48.9	50.3	44.2
• combustible materials	36.3	34.9	32.7	29.0	30.1	32.9
• uncombustible materials	23.4	24.2	24.4	22.1	19.6	23.0
• LHV (as received) (kJ/kg)	6327	5920	5585	4275	4723	5422
Composition (%)						
• Paper and board	25.6	23.1	23.3	18.5	20.4	22.3
• Plastics	8.5	7.3	6.4	6.6	8.3	7.2
• Metals	3.3	3.4	3.0	2.5	2.2	3.0
• Organics	37.4	40.4	40.7	46.3	47.9	42.1
• Inerts	7.6	9.6	8.6	5.0	7.9	7.1
• Unsieved materials	17.7	16.2	18.0	21.1	13.3	18.3

Indagine sui rifiuti solidi urbani in Italia (National Council of Research—Oriented Project for Energy: Survey on Municipal Solid Wastes in Italy)—Rome, April 1980, CNR–PFE LB 3 (see Section 2.3 for the address).

The mean production has been quantified in 0.68 kg/cap/day, with relevant increasing trends and with great regional differences in composition reflecting the habits and the development levels in the various parts of the country. From the Northern to the Southern regions the cellulose and paper percentages show a gradual decrease, while the organic matter value grows increasingly more important. Significant variations are thus observable in the heating value passing from more than 6300 kJ/kg in the North-West to less than 4300 kJ/kg in the South, as well as in the water content (Table 1).

5.0 Municipal Solid Waste (MSW) Collection

The collection of municipal wastes is carried out nearly always by traditional systems, and generally depends on the municipal administration. At present, separate collection activities having a real continuous character are few and concern paper and glass essentially.

Paper is collected by either a door-to-door system or in collection centres, often placed in schools, where the delivery takes place on scheduled days and the transport to the buyers is carried out by the public institutions. The recovery is often greater than 20% of the total paper in MSW in the first case, but generally does not overcome 3–4% in the second one. The use of containers placed in public areas, on the contrary, has been given up, because of the problems arising from abusive collection and acts of vandalism.

Table 2 Some examples of separate collection of paper and glass

	Paper and board collection		Glass collection	
Town	Collection system	Yield (kg/cap/year)	Collection system	Yield (kg/cap/year)
Milan	—	—	containers (M)	6.3
Brescia	school centre	2.17	containers (M)	8.0
Padua	—	—	containers (M)	5.0
Parma	door to door	10.90	containers (S)	5.6
Imola	door to door	14.25	containers (M)	8.5

(M) = mixed colours.
(S) = separate colours.

Container collection represents, however, the usual method for glass separate collection, generally carried out without colour separation, allowing the recovery of about 30–40% of the glass in MSW. The addresses of the main public services attending to separate collection of municipal solid wastes are listed below.

- Azienda Municipale Servizi Ambientali, Tel. 2 25891
  Via Olgettina 25
  20132 MILANO
- Azienda Servizi Municipalizzati, Tel. 30 33111
  Via Lamarmora 230
  25100 BRESCIA
- Azienda Municipalizzata della Nettezza e dell'Igiene Urbana, Tel. 49 776055
  Via Savelli 9
  35129 PADOVA
- Azienda Municipalizzata Nettezza Urbana, Tel. 521 993841
  Viale Piacenza 4/6
  43100 PARMA

6.0 Municipal Solid Waste (MSW) Treatment and Disposal

Up to today, no official statistics exist on the systems in use for municipal solid waste treatment and disposal. After the indicative data reported in Table 3, land disposal appears to be still largely prevailing, even if no comprehensive information is available on sanitary landfills. The indicative value of 73.5% of the refuse disposed in this way was obtained by the difference from the amount disposed by other techniques and thus includes also a number of partially controlled or even free tipping sites. The effec-

Table 3 Treatment and disposal systems for municipal solid wastes in Italy

Treatment and disposal systems	Number of plants	Total capacity (t/day)	% of disposed wastes
Incineration	96	12 833	24
Composting	11	1010	2
Material recovery	2	330	0.5
Landfill	—	—	73.5

tively controlled landfills would be about forty, with a total capacity of 11 450 t/day (see Section 1) and would be used for the disposal of 3.15 million t/year.

Such situations can be chiefly attributed to the remarkable slowing of the construction of new plants in the last years, due to political and public opinion problems relating to environmental impact and—in the case of incineration—to atmospheric pollution.

6.1 Incineration

Comprehensive data on incineration have been collected by the Ministry of Health, by a survey carried out during 1979–1982:

- Giannico, L., Seller, L. (1982): Indagine sugli impianti di incenerimento dei rifiuti solidi urbani in Italia (Survey on incinerators for municipal solid wastes in Italy), Proceedings of SEP Pollution, Padova, 1982, 367–406.

After such work, it turned out that 96 incineration plants were operating in Italy—63 with multistage grates and 21 with rotating drums—with a total capacity of 12 833 t/day and 134 t/day on average for each plant. Incinerated wastes amounted to 3 371 500 t/year, which corresponded to 24% of the whole national production.

The comparison with a previous survey carried out in 1973 pointed out an important decrease of the operating incineration plants (125 at the time including also the units of compost and incineration mixed plants) for the shut down of several small sized units, while the incinerated amounts (2 440 000 t/year in 1973) increased only slightly (Table 4). As a matter of

Table 4 Municipal solid wastes incineration in Italy. Comparison between the situation in 1973 and 1982

Items	1973 survey	1982 survey
Operating incineration plants	125	96
Installed capacity (t/day)	9500	12 833
Mean plant capacity (t/day)	76	134
Average age of plants (years)	2.6	7.0
MSW incinerated (t/year)	2 439 000	3 371 500
Percentage of MSW incinerated (%)	16.8	24.0

fact, the construction of new plants roughly decreased, as proved by the increase of the average age of the operating incineration plants from 2.6 to 7 years (in 1982). Throughout the five years between 1977 and 1982, the incineration capacity of the new installations has been limited to 1760 t/day, with only eight units of capacity higher than 100 t/day, as a result of choices carried out in previous periods.

After 1982, the number of operating incinerators has further decreased, and no new plants have come into operation. All the realized intervention have concerned the existing facilities, often in order to achieve energy recovery or to get a better control on the gas emission to cope with the new, more restrictive regulations that came into force in the meantime.

The present regulations, in fact, impose, all over the national territory, a secondary combustion chamber (post-combustion) in all municipal solid waste incinerators, coping with the following criteria:

- contact time $\geqslant 2$ sec;
- gas velocity in the inflow section 10 m/sec;
- gas temperature $\geqslant 950$°C.

Gas temperature must be over 1200°C when organic chlorine concentrations in the processed waste exceeds 2%. The limits for pollutant concentrations in the gas emissions are set by the regional government regulations: for large plants the required mean values are generally 40 mg/nm^3 for dusts, 50 mg/nm^3 for hydrochloric acid and 5 mg/nm^3 for heavy metals.

6.2 Material recovery and composting

Material recovery from mixed wastes by mechanical sorting techniques has had unimportant developments in recent years. No new plants for the

Table 5 Material recovery and composting plants in Italy (1986)

	Operating		In construction	
Type of plant	Number of plants	Total capacity (t/day)	Number of plants	Total capacity (t/day)
RDF and compost	1	80	7*	950
RDF, compost and incineration	1	250	—	—
Compost and incineration	2	160	4	600
Compost	9*	850	2	150
Total	12	1340	13	1700

*Two plants, presently designed only for composting, are being completed by a line of RDF production and have been included in the table.

separation of specific fractions (paper, glass, plastics, etc.) have been constructed after the experience of Rome and Perugia plants, dating up to the 1970s. The existing plants in Rome are now partly inactivated, due to the problems encountered in the marketing of recovered products, and are only operating for the production of compost allowed by law. Their future is actually in discussion, so that they have not been included in the calculation of data reported in Tables 3 and 5. However, Perugia facilities are still in operation, processing 250 t/day, even if paper recovery is no longer performed. Such plants produce densified RDF, granulated plastics and compost. The composting line is presently being modernized.

The production of RDF, in its different forms, often obtained with compost is the main present trend in Italy. Besides the Perugia plant, only one plant is now operating, processing 80 t/day and producing RDF and composting, but seven are being constructed or transformed, and will receive, on the whole, 950 t/day.

Eleven plants are then working only for compost production (processing a total amount of 1010 t/day). Six more composting plants are being constructed, and will process 750 t/day.

The present law, however, imposes precise constraints for the agricultural use of compost, and the respect of the quality standards reported in Table 6.

Table 6 Quality standards set by the Italian law for compost. All values are expressed as % on dry matter basis

Parameters	Limit values (% d.w.)
Inert materials	≤ 3
Glass	≤ 3
Plastics	≤ 1
Iron	≤ 0.5
Water content	<45
Organic matter	>40
Humidified organic matter	>20
Total nitrogen	>1
P_2O_5	>0.5
K_2O	>0.4

The composting process must include a fermentation period of at least three days at 55°C and the land application rates cannot exceed 30 t/ha over a three year period (only for floriculture some exceptions are allowed). Several constraints have then been set about the application periods and modality for the different kinds of crops. The granulometric class must be included between 0.5 and 25 mm and no glass must be present exceeding 3 mm size. Carbon/nitrogen ratio must be equal to or below 30.

Severe restrictions concern toxic metals; compost cannot be used on soils where metal concentrations exceed the values reported in the first column of Table 7. In all other cases, the amount of compost to be used depends on its metal content, which must respect the limits reported in the third column of the Table 7. Further details on the regulations in force for compost production and use can be found in the text of the regulations cited in Section 8.

Table 7 Constraints for compost use, relating to the presence of heavy metals

Metals	Maximum concentrations in soil to be compost-treated (mg/kg d.w.)	Maximum application rates (g/ha/year)	Maximum concentrations in compost (mg/kg d.w.)
Arsenic	10	100	10
Cadmium	3	15	10
Chromium (III)	50	2000	500
Chromium (VI)	3	15	10
Mercury	2	15	10
Nickel	50	1000	200
Lead	100	500	500
Copper	100	3000	600
Zinc	300	10 000	2500

6.3 Sanitary landfill

Most recent realizations for municipal solid wastes disposal consist in sanitary landfills, often placed in geographic areas where geological structure was such to impose important works of lining and leachate collection. In some Italian regions, especially in the North, this is the most widely adopted system in the last few years. In Lombardy, the regional programme of 1981 provides for the realization of 16 new, large sanitary landfills, designed to receive 2 405 000 t/year.

For landfills lined by artificial materials, the law imposes a number of conditions which can be summarized as follows:

(a) the bottom of the landfill must lay over at least 150 cm from the maximum excursion of aquifer level;

(b) the lining material (usually a HDPE film 2 mm thick) must be applied on an inert layer at least 100 cm thick characterized by a permeability lower than 10^{-6} cm/sec;

(c) leachate must be collected and disposed adequately. Its eventual discharge in surface waters is subjected to the quality standards set by law for industrial waste waters.

The last indication is very restrictive in the case of leachate (BOD = 40 mg/l, COD = 160 mg/l, NH_4^+ = 15 mg/l, SO_4^{2-} = 1000 mg/l, Fe = 2 mg/l, Zn = 0.5 mg/l, etc.). Thus, no individual treatment plant has been constructed. Leachate is recycled on the landfill surface or collected and carried to municipal waste water treatment plants.

The present regulations also provide for biogas collection, in order to control odours and ground infiltrations. Many landfills have been equipped to cope with such prescriptions and burn gas by flares. Energetic use of biogas, on the contrary, is still experimental.

Landfills usually receive raw wastes; two plants only, use baling before tipping (one of them is presently under construction) and no one grinds them.

6.4 Addresses of recent installations

Due to the above-mentioned reasons, the construction of new installations in recent years has been extremely limited. In particular for incinerators, only completion interventions have been carried out. Among the most significant facilities, the following ones can be mentioned:

- Milano (Figino) incineration plants, with two 300 t/day furnaces, superheated steam generation (46 t/h) and turbine capacity of 11.5 MW. They come into operation in 1975 with electrostatic precipitators and were completed in 1984 with wet scrubber and post combustion chambers. Address:
  Azienda Municipale Servizi Ambientali
  Via Olgettina 25
  20132 MILANO

- Valmadrera incineration plant, with two 120 t/day furnaces and steam generation (30 t/h). Came into operation in 1980 with an electrostatic precipitator, and it has successively been equipped with wet scrubber and post combustion chamber. Address:
  Consorzio Intercomunale Eliminazione Rifiuti Solidi
  Via Massena 9
  22049 VALMADRERA (Como)

- Perugia material recovery plants produce densified RDF, granulated plastics, iron and compost (processing 250 t/day). They provide air separation of the lighter fraction and include a bioreactor for composting. Two furnaces (60 t/day each) are designed to receive rejects from

the composting process and generate steam for heating and internal technological use. Came into operation in the 1970s and have been transformed to perform the present treatment cycle in 1985. Address:
Gestione Servizi Nettezza Urbana S.p.A.
Via della Molinella 7
06100 PERUGIA

- Tolmezzo material recovery plant, producing RDF-pellets, compost and iron (processing 80 t/day). Composting is performed by two Dano bioreactors and completed by aerated curing piles. Rejects from composting are used for the production of RDF. Came into operation in 1984. Address:
  Comunità Montana della Carnia
  Via Carnia Libera
  33028 TOLMEZZO (Udine)

Among the several landfills equipped with artificial lining and gas gathering systems, the following can be mentioned:

- Gerenzano landfill, came into operation in the 1970s and can receive a total amount of 12 million m^3; 1.5 millions of volume are still available. The most recent parts have been lined by a HDPE layer. Leachate is collected and recycled in the refuse mass. Biogas is gathered and burnt by flares. A study is in progress for a general energetic recovery project for 3000 nm^3/h. It serves part of Milan municipality and of the surrounding areas. Address:
  Azienda Municipale Servizi Ambientali
  Via Olgettina 25
  20132 MILANO

- Cavenago landfill. Came into operation in 1976, with a global capacity of 2.6 million m^3, it is mostly lined by HDPE layer. Leachate is collected and recycled in the refuse mass, while the excess is sent to external municipal sewage treatment plants. Biogas is gathered and burnt by flares. A study is in progress for energy recovery (500 nm^3/h). Address:
  Consorzio Smaltimento Rifiuti Solidi Urbani
  Commune di Inzago
  20065 INZAGO (Milano)

- Castagneto landfill. It serves the Brescia municipality and surrounding areas. Came into operation in the 1970s, and it is artificially lined by HDPE layer in the more recent zones (global capacity: 2.5 million m^3). Leachate is collected and sent to the Brescia waste water treatment

plant. Biogas is presently burnt by flares but a project exists for its exploitation to produce electric power. Address:
Azienda Servizi Municipalizzati
Via Lamarmora 230
25100 BRESCIA

7.0 Industrial Wastes

The only comprehensive data on industrial waste generation in Italy are derived from a survey carried out per sample. The results were published in 1977 by the Confederazione Generale dell'Industria Italiana (General Confederation of Italian Industry):

- Confederazione Generale dell'Industria Italiana: I rifiuti solidi industriali; elementi conoscitivi e propositivi per una politica di intervento (Industrial solid wastes; cognitive and propositive elements for an intervention policy). Ed. SISPI, 1977.

The manufacturing industry was found to generate at that time 34.9 million t waste/year, subdivided, as shown in Table 8, among the various activities, with a clear prevalence of the metallurgical sector and, to a lesser extent, of building materials, mechanical, foodstuff, means of transport and chemical sectors. These six sectors covered generally more than 88% of the total amount produced. The indicated quantities include—besides inert materials—wastes which were already recovered in the productive cycles by the same generating industry or which were sold to external settling. In

Table 8 Industrial solid waste generation in the different sectors of manufacturing industry (1977)

Industrial sector	Generated amounts (million t/year)	% of the total generated amount
Metallurgy	17.3	49.6
Building materials	4.4	12.6
Mechanics	3.0	8.6
Foodstuff	2.3	6.5
Means of transport	2.0	5.7
Chemistry	1.9	5.4
Paper	0.8	2.3
Leather and hides	0.6	1.7
Others	2.6	7.6
Total	34.9	100

Table 9, a subdivision is reported relating to the main typologies and to the geographic origin. The contribution of Northern regions appears to be largely prevailing (72%) reflecting the higher industrial concentrations.

Table 9 Subdivision of industrial solid wastes for typology and geographic origin (1977). Values expressed as 10^3 t/year

	Geographic area			
	North	Centre	South	Italy
Assimilable to MSW	1287	289	201	1777
Recovered or sold	10 520	1892	1863	14 275
Inert materials	10 340	2066	1933	14 339
Plastic materials	100	19	11	130
Rubber	28	8	7	43
Sludges	2406	662	537	3605
Inflammable residues	61	17	7	85
Others	438	100	111	649
Total	25 180	5053	4670	34 903

Data referring to 1977, however, cannot describe realistically the present situation, due to the transformations which took place in both the productive structure of the country and in the policy of environmental control (increase in sludge generation, new regulations on industrial waste classification).

To fill in the gaps in the present knowledge, a large collection of data is taking place, coordinated by Istituto Superiore di Sanità (see Section 2.2 for the address) concerning six Italian regions (Lombardy, Veneto, Sardinia, Emilia-Romagna, Sicily and Campania). They have been chosen on the basis of their different economic and social conditions. The end of the survey is foreseen for 1988. Another survey, including wastes of agricultural and zootechnic origin is in progress within the frame of the oriented project for energy (2) (see Section 2.4 for address).

The collection and disposal of industrial wastes is subjected to the recent regulations (see Section 8). All the matter is still evolving, and few centres only, nearly always depending on private companies, carry out regular activities in the field of treatment and disposal. The main centres are listed below, the first of which is the only public managed one:

(a) Azienda Municipalizzata Igiene Urbana, Tel. 59 372430–372464
Via Morandi 54
41100 MODENA

It is the public cleansing organization operating in Modena. Municipal solid wastes are disposed of by a 300 t/day furnace and by a 950 000 m^3 sanitary landfill. A 150 t/day fixation plant is working for inorganic sludges (Soliroc system) and a treatment plant (20 m^3/day) receives oil wastes. A 36 t/day incinerator is being built for toxic and hazardous wastes.

(b) Ecodeco S.p.A.
Cassinazza di Baselica
27010 GIUSSAGO (Pavia)

The company has a series of facilities, in the surroundings of Milan, including:

- incineration plant for liquid wastes (30 000 t/year), with wet scrubber and post combustion chamber, performing electric power recovery (0.5 MW);
- fixation plant for inorganic sludges (Ecodeco system), with a 200 000 m^3 landfill;
- treatment centre for biological sludges and agricultural recovery of industrial wastewaters (100 000 t/year).

(c) Ecoservizi S.p.A., Tel. 30 41596–296483
Via dei Santi 58
25010 BETTOLE (Brescia)

The company holds a biological and a physicochemical treatment plant for industrial wastewaters and a 50 000 t/year fixation plant (Ecoservizi systems), with a 150 000 m^3 landfill.

8.0 Legislation

The basic text of Italian law concerning wastes is represented by:

- Decreto del Presidente della Repubblica, settembre 10, 1982, n. 915: Attuazione delle Direttive CEE n. 75/442 relativa ai rifiuti, n. 76/403 relativa allo smaltimento dei policlorobifenili e policlorotrifenili e n. 78/319 relativa ai rifiuti tossici e nocivi (Application of the EEC Directives n. 75/442 concerning wastes, n. 76/403 concerning the disposal of polychlorinated biphenyls and triphenyls and n. 78/319 concerning toxic and hazardous wastes), published on Gazzetta Ufficiale della Repubblica Italiana n. 343 of December 15, 1985.

The decree sets the general principles for waste collection, transport and disposal, performs a classification of wastes (municipal, special, toxic and hazardous), defines the competence of the central administration, of the local governments (regions and districts) and of the municipalities, indicates the conditions and procedures for the various phases of the authorization to toxic and hazardous waste disposal.

The national technical directives, cited in the previous paragraphs, are included in:

- Deliberazione del Comitato Interministeriale, luglio 27, 1984: Disposizioni per la prima applicazione dell'art. 4 del Decreto del Presidente della Repubblica settembre 10, 1982, n. 915, concernente lo smaltimento dei rifiuti (Provisions for the first application of thc art. 4th of the Decree of the President of the Italian Republic of September 10, 1982, n. 915, concerning waste disposal), published on Supplemento Ordinario della Gazzetta Ufficiale della Repubblica Italiana n. 253, settembre 13, 1984.

Such provisions detail the criteria for the classification of special, toxic and hazardous wastes; provides technical indications for waste treatment plants, set the quality standards for recovered products and the parameters to comply with for the construction of landfills and indicate the general criteria for the authorization to dispose toxic and hazardous wastes as well as the methods for sampling and analysis.

The organization of disposal services and the procedures for control and authorization are the object of technical provisions issued by several local governments to integrate the central government law. For a global picture of waste legislation, see:

- Giampietro, F., Giampietro, P. Lo smaltimento dei rifiuti—Commento al DPR 915–1982 (Waste Disposal—Comments on DPR 915–1982). Maggioli Publishing Company, Rimini, 1985.

The Gazzetta Ufficiale della Repubblica Italiana and all the other official publications can be obtained from:

- Agenzia dell'Istituto Poligrafico e Zecca dello Stato, Tel. 6 85082149–85082227.
  Piazza G. Verdi 10
  00198 ROMA

9

Japan*

Surface: 369 640 km^2

1.0 General

1.1 Enactment and amendment of "Waste Disposal and Public Cleansing Law"

In Japan, waste management is conducted under the "Waste Disposal and Public Cleansing Law" (also referred to as "Waste Disposal Law").

In the past, waste management was conducted as treatment of filth in the special cleansing area under the Public Cleansing Law. But with intensive development of economics and improvement of living, wastes have increased greatly in quantity and changed in quality. In this situation, the Public Cleansing Law was amended entirely and substituted by the Waste Disposal Law in 1970. The law aims to consolidate waste management schemes to cope with the situation, preserve the living environment and improve the public health. In 1976, the Waste Disposal Law was amended to secure the proper management of industrial wastes.

1.2 Outline of waste management system

Under the Waste Disposal Law, wastes are classified into two categories: "Domestic Wastes" and "Industrial Wastes".

Industrial wastes are wastes which are discharged in connection with business activities and must be disposed appropriately on the responsibility of enterprisers. Domestic wastes are wastes other than industrial wastes, including night soil, which are discharged mainly from households.

The responsibility of the disposal of domestic wastes is placed on the respective municipalities. The municipalities set forth a definite programme on the management of domestic wastes in the respective area, as they collect, transport, treat and dispose domestic wastes in accordance with the programme.

There are three methods for collection, transportation, treatment and

---

* By Mr. Tadayuki Morishita and Dr. Katsumi Yamamura.

disposal of domestic wastes. The municipalities directly conduct the principal method. Other than that, commissioned undertakers conduct as a substitute for the municipalities, or licensed domestic wastes contractors do so. Domestic wastes must be collected, transported, treated and disposed in accordance with the standards specified in the cabinet order not to interfere living environment.

As for industrial wastes, it is the responsibility of the enterprises connected. Industrial wastes are managed by enterprises themselves or by the contractor permitted by prefectural government.

The prefectural government endeavours to control industrial waste management in the respective prefecture along with provisions if necessary, and decides on a master plan for the management of industrial wastes. They provide enterprises with guidelines of management and they lead enterprises.

The national government promotes technical development on waste disposal along with the endeavour to give necessary technical and financial assistance to the municipalities and the prefectural governments.

For the industrial wastes containing hazardous substances, more strict management standards are specified.

1.3 Present problems to be solved

Seventeen years after the enforcement of the Waste Disposal Law, waste management rises in importance. At present, the following problems are:

- With improvements in people's living standards, there is an increasing proportion of the wastes that is hardly disposable in the waste management system of the municipalities, for example, waste home electric appliances, construction waste materials and waste plastics.
- To promote the appropriate waste management, it is necessary to increase waste treatment facilities and final disposal sites. However, it has become very difficult to obtain the consensus of inhabitants for construction of such facilities. In some cases, local means such as roads and water supply facilities are constructed around the site of waste treatment facilities to obtain the consensus. It is necessary to examine how to evaluate these constructions in the waste management systems and to carry them out efficiently.
- In some cases, it becomes difficult for the enterprisers and licensed contractors to manage industrial wastes. So it is necessary for the public organization to participate in the treatment of industrial wastes. Many local public corporations for industrial waste treatment are organized in participation of the prefectural government and big cities.
- So far as the difficulty of obtaining final disposal sites and national

demands of saving resources and energy is concerned, resource recovery of wastes becomes a great concern. Various measures have been planned to promote resource recovery such as the introduction of source separation for recovering useful materials, the development of techniques for separation from collected wastes, generating electric power and/or providing heat in incineration plants, etc.

- Scattered wastes, especially by soft drink's cans, are also a big problem to be solved urgently.

For these problems, it is necessary to make long-term plans.

1.4 Current main topics

The 6th five-year national program for waste treatment and disposal (1986–1990) was determined in the cabinet council on April 25, 1986, to promote urgent and nationwide construction of the waste treatment and disposal facilities. In accordance with this programme, ¥1910 billion will be invested for waste treatment facilities (including nightsoil treatment) in five years. The ratio of the wastes which are treated in the incinerator to the whole combustible wastes will be raised from 88% to 92% (at March 1987).

As mentioned before, it is very difficult to obtain final disposal sites. To cope with this situation, there are plans to promote the regional waste disposal programme. This programme's object is to set final disposal sites on the seacoast in cooperation with more than two prefectures including many municipalities. To facilitate, a new law was promulgated in 1981. (See Section 4.)

Pneumatic transportation systems are now examined instead of present collection and transportation systems operated with vehicles. Ministry of Health and Welfare and Ministry of Construction selected several cities as model trials for introducing a pneumatic transportation system and in two cities, this system has begun to operate. Although the costs are high at present and there are many difficulties to solve generally, it is expected to be evaluated as one of the directions of waste management systems in future.

2.0 Status of Domestic Wastes

2.1 Discharged quantity of domestic wastes

The discharged quantity of domestic waste (domestic solid wastes) is stated to be closely correlated with the gross national product (GNP). This means that from 1955 to 1970, along with the activation of economical activity, the discharged quantity has increased and with the oil crisis in 1973 as the borderline, its growth has stopped. However, the daily discharge per person differs with respective cities, as shown in Table 1. Generally, the people living in large cities generate more domestic wastes than those in small cities.

Table 1 Daily discharge per person in various cities (FY 1979)

City	Population (thousand)	Discharge (g/day person)
Sapporo	1240	704
Tokyo	8647	1675
Kawasaki	1015	991
Nagoya	2088	1314
Kyoto	1461	1165
Osaka	2779	1628
Fukuoka	1058	941

2.2 Domestic waste disposal flow

The flow of domestic wastes is shown in Tables 2 and 3 according to the annual survey on waste management by Ministry of Health and Welfare (MHW).

The average daily discharge of domestic wastes in municipal management areas was 100 066 t in FY 1984. Of this, 94 740 t were collected by municipalities and the remains were brought into treatment or final disposal facilities directly by the dischargers themselves.

Domestic wastes are incinerated for the purpose of reducing their volume and stabilizing them. Residues are then disposed into landfill sites. As of the end of FY 1984, domestic wastes treated by incineration constituted about 72.1%, and wastes disposed directly into landfill sites constituted about 19.6%. In Table 4 the schemes for collecting the domestic wastes are shown. They vary from city-to-city due to differences in the historical and social backgrounds of the respective cities.

Table 2 Changes of refuse treatment (whole country)

Fiscal year	1980	1981	1982	1983	1984
Per capita generation (g/day)	809	828	842	826	833
MSW generation from household (t/day) : (figs. in %)	94 354 (100.0)	97 418 (100.0)	99 831 (100.0)	98 417 (100.0)	100 066 (100.0)
(1) Collected by municipality and disposed of by:					
• Incineration	63 177 (67.0)	65 487 (67.2)	69 105 (69.3)	69 491 (70.6)	72 161 (72.1)
• Landfill	21 929 (23.2)	22 040 (22.6)	21 559 (21.6)	20 108 (20.5)	19 557 (19.6)
• High speed composting	198 (0.2)	94 (0.1)	117 (0.1)	140 (0.1)	130 (0.1)
• Composting and animal feeds	76 (0.0)	42 (0.0)	43 (0.0)	61 (0.0)	52 (0.0)
• Others	2331 (2.5)	3146 (3.2)	2406 (2.4)	2729 (2.8)	2840 (2.9)
• Total	87 711 (93.0)	90 809 (93.4)	93 230 (93.4)	92 529 (94.0)	94 740 (94.7)
(2) Not collected and disposed at source	6643 (7.0)	6609 (6.8)	6601 (6.6)	5888 (6.0)	5326 (5.3)
MSW from other sources disposed at municipal facilities	26 017	19 400	22 026	18 447	17 850

Table 3 Domestic wastes disposal flow (t/day)

Treatment and disposal methods	MSW collected directly by municipality 91 593	Bulky wastes separately 3146	MSW received and disposed at municipal facilities 17 851	Total 112 590
Crushing and/or compaction	—	1609	—	1609[a]
Animal feeds	52	—	20	72
High speed composting	130	—	4	134[b]
Incineration	70 939	345	5679	77 840[c]
Others	2566	86	168	3008[d]
Landfill	17 906	1106	11 980	44 372[e]

[a] After crushing and/or compaction, wastes are disposed as follows: 544 t to landfill, 877 t to incineration, 188 t by other ways.

[b] 29 t/day of residues from composting are landfilled.

[c] The total amount also includes the amount coming from crushing and/or compaction (see [a]). 12 363 t/day of slags and ashes from incineration are landfilled.

[d] 444 t/day of residues from other treatments are landfilled.

[e] The total amount also includes the amounts coming from crushing and/or compaction (see [a]), the residues from composting (see [b]), ashes and slags from incineration (see [c]) and residues from other treatments (see [d]).

Table 4 Status of refuse collection (whole country, fiscal year 1984)

		t/day
Government (city, town, village)	Direct management	53 791
	Contract	24 381
By approved contractors		13 736
Total		91 908

2.3 Physical composition of domestic wastes

Examples of physical composition in domestic wastes are shown in Table 5. The proportion of combustibles is around 70% in typical domestic waste in Japan. Water content of domestic wastes is about 50%, though it is tending to decrease recently. Typical lower calorific values of domestic wastes in Japan is between 1300–2000 kcal/kg, while the values vary with the change of physical composition and water content. The lower calorific value has increased recently due to the decrease of water content and the increase of plastics content.

Table 5 Physical composition of domestic wastes

	Composition	Air dry base (%)	
		Kawasaki (1979)	Osaka (1980)
Combustibles	Papers	35.3	37.1
	Feed residuals	13.5	13.2
	Straw, grass	2.9	2.5
	Textiles	3.0	4.0
	Wood splits	3.6	2.5
	Leathers, rubbers	0.5	0.3
	Plastics	9.7	15.2
	Others	0.1	—
	Subtotal	68.6	74.8
Incombustibles	Shells, bones	1.5	1.4
	Sands, stones	2.8	6.0
	Metals	7.5	5.5
	Others	19.6	12.3
	Subtotal	31.4	25.2

2.4 Waste treatment facilities

At the end of fiscal year 1984, the capacity of the waste incineration facilities had increased by 2.8%. The number of facilities for the treatment of bulky wastes at the end of 1984 is 494. The high speed composting facilities number 28 with a treatment capacity of 598 t/day.

Table 6 Status of waste incineration facilities (on the basis of initiation of the construction work—whole country

Fiscal year	Number of facilities	Total design capacity (t/day)
1980	2011	147 394
1981	1998	148 792
1982	1962	151 777
1983	1938	153 692
1984	1916	158 071

The landfill sites are shown in Table 7. Total landfill capacity remains at about 174 million m^3. However this corresponds to the capacity for only 5–6 years in the future.

Table 7 Status of landfill sites

Fiscal year	Number of landfill sites location					Total area ($\times 10^3 m^2$)	Total volume ($\times 10^3 m^3$)	Remaining volume ($\times 10^3 m^3$)
	Mountainous	Shore	Ponds, marshes, swamps, etc.	Flat area	Total			
1980	1600	36	50	796	2482	52 086	356 109	191 945
1981	1619	40	48	779	2486	53 581	403 156	181 578
1982	1612	36	46	778	2472	53 929	377 583	175 975
1983	1638	34	41	766	2479	55 094	382 728	170 795
1984	1633	37	38	731	2439	56 246	403 062	174 216

2.5 Expenses for waste disposal works

Under the Waste Disposal Law, national government is required to give financial aid to municipalities in the construction of waste treatment facilities. National government subsidizes one-quarter to half of the expenses for plant construction, and the rest is covered by municipal loans guaranteed by national government.

The expenditure of municipalities for domestic waste management was more than 1000 billion yen in FY 1984, which shares about 5% of the total expenditure of municipalities.

The expenditure required for waste treatment businesses has rapidly increased, and during the last ten years, it has expanded about ten times. The national financial assistance in terms of government subsidies and grants has also rapidly increased.

Average annual management cost of domestic wastes in Japan was 8030 yen/person in 1979. However, the management cost differs with respective municipalities, because it depends on the collection frequency, transportation distances, treatment method etc. However, the management cost per ton of collected domestic waste was 23 501 yen as a national average. About 70% of the total cost are due to collection and transportation, which is more expensive in large cities because of traffic conditions.

3.0 Status of Industrial Wastes

3.1 Present status of industrial wastes treatment

Industrial wastes are those discharged in connection with business activities. They are as follows:

- Cinder
- Sludge
- Waste oil
- Waste acid
- Waste alkali
- Waste plastics
- Waste paper
- Wood chips
- Waste rags
- Vegetable and animal remains
- Rubber scraps
- Metal scraps
- Glass and porcelain chips
- Slags

- Construction waste materials
- Livestock excretions
- Livestock corpses
- Dust
- Substances treated from industrial wastes for the purpose of disposal

The volume of industrial wastes discharged annually in Japan is estimated at 292 million tonnes. Slags account for the largest portion, about 61 million tonnes, or 20.7% of the total (Table 8). It is followed by livestock excretions

Table 8 Discharges of industrial wastes throughout the country (1975)

Type of industrial waste	Annual discharge (1000 t/year)	Discharge percentage (%)
Cinder	1797	0.6
Sludge	88 190	30.2
Waste oil	2419	0.8
Waste acid	10 219	3.5
Waste alkali	6090	2.1
Waste plastics	2232	0.8
Rubber scrap	92	0.0
Metal scrap	13 111	4.5
Glass and porcelain chips	2297	0.8
Vegetable and animal remains	4323	1.5
Waste paper	1624	0.6
Wood chips	6628	2.3
Waste rags	101	0.0
Slags	60 561	20.7
Construction waste materials	30 007	10.3
Livestock excretions	49 629	17.0
Livestock corpses	62	0.0
Dust	11 731	4.0
Treated waste	1199	0.4
Total	293 312	100.0

(50 million tonnes), sludge (88 million tonnes) and construction waste materials (30 million tonnes). Each of these exceeds 10% of the total. These top four categories account for 78.2% of the total volume of discharge. Industry, iron, agriculture and construction together account for a total of 49.8%. The combined share of these three and the ceramic, clay and stone products manufacturing and mining industries amounts to about two-thirds of the total volume of industrial wastes discharged (Table 9).

Table 9 Waste quantities classified by industries (%)

Type of industry	Waste production (%)
Iron and steel industry	22.3
Agriculture	17.1
Mining	12.7
Construction	10.4
Ceramics	6.3
Food and tobacco industry	4.7
Pulp, paper and paper processing industry	4.7
Chemical industry	3.6
Nonferrous metal industry	3.3
Electricity, city gas, city water and heat supply	2.8
Others	12.1

With respect to the relationship between the discharges of individual categories of wastes and the industries, 81.3% of the volume of slag discharged is accounted for by the iron industry and 13.4% by the mining

Table 10 Relation between waste quantities classified by type and industries (1975) (%)

Industries producing waste	Type of wastes			
	Slags	Livestock excretions	Sludge	Construction waste materials
Iron industries	81.3	—	—	—
Mining	13.4	—	11.7	—
Agriculture	—	100	—	—
Ceramic, clay and stone	—	—	17.4	—
Pulp, paper and paper processing	—	—	15.4	—
Electricity, city gas, city water supply, sewage system and heat supply	—	—	13.1	—
Chemical industry	—	—	11.3	—
Construction industry	—	—	—	90.2
Others	5.3	—	31.1	9.8
Total	100 ($\sim 61 \times 10^6$ t)	100 ($\sim 41 \times 10^6$ t)	100 ($\sim 38 \times 10^6$ t)	100 ($\sim 34 \times 10^6$ t)

industry. This means that most of the slag is discharged by these two industries. Likewise, most of the livestock excretions and of the construction waste materials are discharged by the agricultural and the construction industry, respectively.

The sludge is discharged not by a few specific industries but by a wide variety of industries (see Tables 10 and 11). 53% of all the industrial wastes discharged undergo intermediate treatment while 50 million tonnes are directly turned to final disposal (1980). At the same time, 88 million tonnes of industrial wastes are recycled for use. Of the volume of industrial wastes directly disposed of, about 6.5 million tonnes are dumped into the sea. Therefore, it is presumed that about 61.5 million tonnes are landfilled. This is about four times as much as the landfill of domestic wastes.

Table 11 Relation between waste quantities classified by industry and kind of waste (1975) (%)

Type of waste	Iron	Agriculture	Construction	Ceramic, clay and stone products	Mining
Slag	77.3	—	—	—	65.0
Dust	6.1	—	—	—	—
Waste acid	5.2	—	—	—	—
Metal scrap	4.8	—	—	—	—
Livestock excretions	—	99.7	—	—	—
Construction waste material	—	—	91.7	—	—
Sludge	—	—	5.5	51.7	35.0
Waste alkali	—	—	—	23.3	—
Others	6.6	0.3	2.8	25.0	—
Total	100 ($\sim 64 \times 10^6$ t)	100 ($\sim 41 \times 10^6$ t)	100 ($\sim 34 \times 10^6$ t)	100 ($\sim 13 \times 10^6$ t)	100 ($\sim 13 \times 10^6$ t)

3.2 Present situation of industrial wastes

Treatment of the industrial wastes is also performed by the enterprisers themselves, the licensed contractors and performed over a regional area by the prefectures. The case of licence granted to the contractors for management of industrial wastes as of 1981 are as shown in Table 12.

Table 12 Number of licensed contractors for management of industrial wastes (as of 1981)

	Total	Collection and transportation only	Intermediate treatment only	Final disposal only	Collection, transportation and intermediate treatment	Collection, transportation and final disposal	Intermediate treatment and final disposal	Collection, transportation, intermediate treatment and final disposal
Number of agencies	26 188	23 226	264	171	1393	843	41	250
Percentage (%)	100.0	88.7	1.0	0.6	5.3	3.2	0.2	1.0

Nowadays, however, difficult problems are occurring in the management of industrial wastes by the enterprisers or the contractors. There are increasing cases of the local public corporations performing the treatment of industrial wastes directly or indirectly from the point of view of conserving the living environment. From June 1986, the industrial waste disposal by public corporations is carried out at 22 prefectures and ten cities.

In view of such a situation, the Ministry of Health and Welfare is specifically examining what should be the public's participation in the management of industrial wastes.

3.3 Future strategies of waste treatment

The management system of industrial waste is based on the principle of the enterprisers being responsible for their own treatment. But, nowadays, there are strict controls of ocean dumping by "Convention on the Prevention of Marine Pollution by Dumping of Wastes and Other Matter" and difficulties in securing disposal sites etc. Thus, in some cases, it is difficult for the enterprisers to carry out the responsibility alone. In the Ministry of Health and Welfare, information is collected concerning the generation and treatment of industrial wastes, developments on waste treatment technology, education and training carried out and therefore the foundations for the treatment of industrial wastes are laid. It also has the benefit of instructing the enterpriser and the contractor to decrease industrial waste discharges and promote the recycling and proper intermediate treatment.

The Ministry of International Trade and Industry has also been taking comprehensive measures in the field of technological research development, etc., in order to promote the reutilization of industrial wastes.

4.0 Regional Planning for Waste Disposal

4.1 Status and problems on waste management in and around big cities

Table 13 shows the amount of generation of wastes in FY 1980, where the Tokyo, Osaka and Nagoya metropolitan regions are concerned.

The area of these three regions is only occupied with 24% of the whole country, but the domestic wastes of 26 million tonnes (ca. 63% of the whole) and industrial wastes of 131 million tonnes (ca. 45% of the whole) have been generated in these regions.

Most of the discharged wastes are reduced by the intermediate treatment such as incineration, dewatering etc. but the residual wastes are ultimately disposed by landfill. Further nonintermediate treated wastes are directly

Table 13 Amount of waste in and around big cities (million t/year)

	Tokyo metropolitan region	Osaka metropolitan region	Nagoya metropolitan region
Industrial wastes	55	38	13
Domestic wastes	12	7	4
Total	67	45	17

disposed to landfill. For proper waste management, final disposal is an essential element and obtaining a final disposal site has become a very important administrative task. In particular, in-and-around big cities, acute difficulties are encountered when performing this task because the land has been utilized too intensively to accept the discharged amount. A serious hindrance will occur with promoting proper treatment of wastes unless some new counter-measure is taken for this problem.

The present situation of final disposal of domestic wastes is shown by the percentage of the population disposing their wastes within their administrative boundaries. This figure only counts 52% in the Tokyo metropolitan region and 53% in the Osaka metropolitan region. It becomes lower as municipalities move into a higher population density. Future percentage estimates are 28% in the Tokyo metropolitan region and 41% in the Osaka metropolitan region; these figures are considerably lower than the present percentage. The tendency with population density will be continued in the future. (Tables 14 and 15.)

Because of the above-mentioned reasons, it is obvious that securing the final disposal site within the local administrative districts will become more difficult in these regions. Many municipalities will not be able to obtain the final disposal site within their own administrative districts in the future and the number of these municipalities will increase expeditiously year by year.

4.2 Plan for securing the final disposal site in and around big cities

When taking a view of the future of waste management in these regions, waste reduction at the source, development of an intermediate treatment and promotion of reuse and recovery, these objectives should be directed with community efforts as much as possible. It is estimated that the discharging amount of wastes to be disposed by landfill may carry on increasing. On the other hand, many municipalities cannot have any prospects of getting the final disposal sites within their administrative districts. So

Table 14 Present situation of final disposal of domestic wastes

Tokyo metropolitan region (numbers in parentheses represent population 10^3 cap. (%))

Situation / Population density (cap/km^2)	Own facilities: Within district	Own facilities: Seacoast	Own facilities: Outside	Intermunicipal organization's facilities	Other municipality's facilities	Approved contractor's facilities	No answer	Total
0–2500	6098 (79)	—	224 (3)	986 (13)	131 (2)	304 (4)	10 (100)	7753 (100)
2500–4500	4343 (92)	—	—	249 (5)	34 (1)	22 (0)	58 (1)	4706 (100)
4500–6500	4885 (80)	—	—	227 (3)	123 (2)	794 (13)	87 (1)	6116 (100)
6500–	171 (2)	9271 (85)	—	666 (6)	81 (1)	773 (7)	—	10 962 (100)
Total	15 497 (52)	9271 (31)	224 (1)	2128 (7)	369 (1)	1893 (6)	155 (1)	29 537 (100)

Table 14—Contd.

Osaka Metropolitan Region

Situation / Population density (cap/km^2)	Own facilities			Intermunicipal organization's facilities	Other municipality's facilities	Approved contractor's facilities	Others	No answer	Total
	Within district	Seacoast	Outside						
0–2500	5091 (65)	—	64 (1)	1235 (16)	797 (10)	39 (1)	256 (3)	305 (4)	7787 (100)
2500–4500	654 (45)	—	60 (4)	161 (11)	456 (31)	—	25 (2)	107 (7)	1463 (100)
4500–6500	778 (36)	—	243 (11)	48 (2)	837 (39)	17 (1)	—	244 (11)	2167 (100)
6500–	—	2625 (59)	—	529 (12)	788 (18)	—	526 (12)	—	4468 (100)
Total	6523 (41)	2625 (17)	367 (2)	1973 (12)	2878 (18)	56 (0)	807 (5)	656 (4)	15 885 (100)

Table 15 Future situation of final disposal of domestic wastes

Tokyo metropolitan region (numbers in parentheses represents population 10^3 cap. (%))

Situation / Population density (cap/km^2)	Own facilities: Within district	Own facilities: Seacoast	Own facilities: Outside	Intermunicipal organization's facilities	Other municipality's facilities	Approved contractor's facilities	Others	No answer	Total
0–2500	5141 (66)	—	218 (3)	1296 (17)	46 (1)	195 (3)	556 (8)	301 (4)	7753 (100)
2500–4500	2497 (53)	700 (15)	—	285 (6)	501 (11)	—	506 (11)	218 (5)	4707 (100)
4500–6500	394 (6)	2686 (44)	411 (7)	630 (10)	538 (9)	691 (11)	555 (9)	211 (3)	6116 (100)
6500–	118 (1)	—	71 (1)	395 (4)	333 (3)	57 (1)	1236 (11)	8750 (80)	10 960 (100)
Total	8150 (28)	3386 (11)	700 (2)	2606 (9)	1418 (5)	943 (3)	2853 (9)	9480 (32)	29 536 (100)

Table 15—Contd.

Osaka Metropolitan Region

Situation / Population density (cap/km^2)	Own facilities			Intermunicipal organization's facilities	Other municipality's facilities	Approved contractor's facilities	No answer	Total
	Within district	Seacoast	Outside					
0–2500	6246 (80)	293 (4)	98 (1)	769 (10)	186 (2)	146 (2)	48 (1)	7786 (100)
2500–4500	551 (38)	391 (27)	—	273 (19)	26 (2)	222 (15)	—	1463 (100)
4500–6500	1477 (68)	73 (3)	—	293 (14)	94 (4)	232 (11)	—	2169 (100)
6500–	67 (1)	3327 (74)	170 (4)	—	—	904 (20)	—	4468 (100)
Total	8341 (53)	4084 (26)	268 (2)	1335 (9)	306 (2)	1504 (9)	48 (0)	15 886 (100)

in future, we should consider integrated regional planning beyond municipalities' boundaries or prefectures' boundaries. With respect to a regional disposal site for landfill corresponding to a massive final disposal demand, it is almost impossible to expect to obtain enough space in an inland area, because land utilization in the metropolitan regions will be promoted with high speed. Urbanization and industrialization needs wide areas for housing and high-productive land use. This means that these developments reduce the possibility for waste landfill use. Though suitable land of low productivity such as poor forest and marshy ground could be found to some extent in rural areas, it may be located far from the waste generating point. Therefore waste transport will be too expensive. So a disposal space for regional landfill can not be found except on the seacoast. Locating disposal sites on the seacoast has the advantage that the depth of landfill wastes is much greater than that of inland. Even with the seacoast, however, there is a limit to the remaining appropriate sites. For utilization of the seacoast and preserving the environment, a large disposal plan needs to be considered under an integral aspect, arranging the opinions and interests of the local public bodies concerned.

5.0 References

- Mr. Tadayuki Morishita
  Country Report on the Status of Solid Waste Management, The Seventh Japan–United States Governmental Conference on Solid Waste Management, June 17–18, 1987, Tokyo, Japan, pp. 17–39.
- Dr. Katsumi Yamamura
  Country Report on the Status of Solid Waste Management, The Fifth Japan–United States Governmental Conference on Solid Waste Management, September 28–29, 1982, Tokyo, Japan, pp. 18–48.

10

The Netherlands

Surface: 33 000 km^2 of which 3000 km^2 is water.
Municipalities: 714 of which:

- 3 with more than 500 000 inhabitants
- 14 with 100 000 to 500 000 inhabitants
- 36 with 50 000 to 100 000 inhabitants
- 126 with 20 000 to 50 000 inhabitants
- 189 with 10 000 to 20 000 inhabitants
- 195 with 5000 to 10 000 inhabitants
- 151 with less than 5000 inhabitants

- The Netherlands is a small country with a very high population density (around 400 inhabitants/km^2).
- "Netherlands" means low country, and the country is very low situated indeed. It is, in fact, a large delta-area of three rivers: the Rhine, the Meuse and the Schelde; a great part of the country is below sea-level.
- The level of the subsoil-water is in general less than 1 m below ground-level. No contact whatsoever is tolerated between waste and subsoil-water.

1.0 Waste Policy

In 1984 the first indicative multiyear programme for waste substances was presented to the Parliament, by the Minister of Housing, Physical Planning and Environment.

The general conclusions in this programme are the following:

- When the Waste Substance Act was drafted and became law, it was assumed that it would be sufficient to control certain waste flows which constituted a problem in much of the country. In view of developments in recent years, however, it is necessary to bring more waste under control. In the years ahead, the emphasis in policy and research will increasingly shift from domestic waste to industrial waste, which greatly exceeds domestic waste in quantity.
- It would hardly be feasible for the authorities to fully control all waste flows, nor is it intended that they should. Priorities will be established

for the extent of government planning per waste flow, on the basis of criteria.

- The closure and clearance of the once-numerous waste dumps must be continued. In the years ahead, such measures must not be confined to waste dumps but must also extend to hospital waste disposal, scrapped motor vehicles, incineration of waste and preparation of compost.
- It will be more effective to prevent the generation of industrial waste than that of domestic waste. This is true in terms of quantity, composition and environmental impact. Measures will be continued with a view to promoting activities by industry to improve processes and so reduce or eliminate environmental pollution and the production of waste and to promote reuse of waste either at the plant where it is produced or elsewhere.
- Separation of waste at source has great potential. Methods of collection (pick-up, delivery or combinations of the two) must be decided on the basis of past evaluation and present research. No recommendations can yet be made.
- The development of mechanical separation methods will be continued. Further research will be performed into ways of simplifying equipment, increasing its reliability, reducing energy consumption and developing uses and markets for waste products.
- Consultations will be held with all concerned with a view to adjust (or further adjust) technical norms relating to quality laid down in the form of regulations, rules for particular sectors of industry and specifications for commodity users, so that waste substances can achieve their full potential as secondary materials or additives. This is a matter of incorporating waste substances into the traditional pattern of commodities and consumable supplies while ensuring that these new substitutes possess the required quality for the application in question.
- The environmental protection standards required for certain applications of waste will not generally be based on waste substances legislation but on other legislation such as the future Soil Protection Act. This will particularly apply to bulk waste for which there is little use outside the construction industry and the construction materials industry, where they can serve as substitutes for surface minerals. Environmental protection standards will be drafted in conjunction with the technical norms referred to in recommendations above. Consideration must be given to what effects on soil and groundwater it is acceptable for waste to have, when it is recycled in this way.
- Recycling and reuse of waste substances can be promoted by encouraging the application of the norms published. Central government should set a good example. Recycling can also be promoted by the waste recovery sector, which has been in existence for a long time, using its

experience of national and international markets to extend its activities to such new categories as construction and demolition waste and waste glass, rubber and plastics.

- The aim for the future in waste handling in The Netherlands is:
  Reuse what is reusable.
  Compost what is compostable.
  Burn what is burnable . . . and tip controlled the rest.

2.0 Addresses

2.1 ISWA national member

Nederlandse Vereiniging voor Reinigingsdirekteuren (NVRD) (Dutch Institute for waste management).

- President:
  Ing. J. P. V. M. Laurijssens
  Postbus 348
  3300 AH DORDRECHT
  Tel. 0 78 163366

- Secretary-General:
  Ing. D Louwman
  Postbus 201
  1620 AE HOORN
  Tel. 0 2290 41444

- Vice-President ISWA:
  Ir. J. A. den dulk
  Frans Halsstraat 3
  5691 EB SON
  Tel. 0 4990 72598

- Scientific and technical committee member:
  Ing. D. Louwman (see above)

- Working group sanitary landfilling: Chairman:
  Ing. D. Louwman (see above)
  Secretary:
  Ir. D. Beker
  R.I.V.M.
  Postbus 1
  3720 BA BILTHOVEN
  Tel. (0)30–49111

- Working Group Hazardous Wastes member:
  Ir. D. den Ouden
  A. V. R. Chemie N.V.
  Postbus 1120
  3180 AC ROZENBURG
  Tel. (0)1819–42911

- Member Editorial Board W.M.R.—Journal:
  Ing. J. Th. Albers
  Schroefstraat 2
  8223 ED LELYSTAD
  Tel. 0 3200 44194

NVRD members are directors, managers and staff-members of waste collection and waste disposal organizations. There are members at municipality level, at regional, provincial and governmental level. There are also members from the private sector.

Since 1970, NRVD runs a training centre for cadre-functions in the waste management field, called: "Uitvoeringscommissie Opleiding Personeel Openbare Reinigingsbedrijven" (OPOR)

Secretary:
Ing. K. Robaard
Turfkade 15
7602 PA ALMELO

2.2 Government offices

- Ministerie van Volkshuisvesting, Ruimtelijke Ordening en Milieubeheer (Ministry of Housing, Physical Planning and Environment)
  Minister:
  Dr. Ed Nijpels
  P.O. Box 450
  2260 MB LEIDSCHEDNAM
  Tel. 0 70 209367

2.3 Addresses in general

- Rijksinstituut voor Volksgezondheid en Milieu (National Institute of Public Health and Environmental Hygiene)
  Director-General:
  Ir. Dr. R. B. J. C. van Noort
  P.O. Box 1
  3720 BILTHOVEN

The RIVM, as a laboratory organization working on behalf of the central government, has as its duty: the recognition, detention, combatting and prevention of dangers which—as far as they can be determined by biological, physical, chemical or technological means—threaten man's health or the quality of the environment.

- Rijksinstituut voor Zuivering van Afvalwater (RIZA) (Institute for Waste Water Purification)
  P.O. Box 17
  AA LELYSTAD

The national institute RIZA is a scientific and technical research institute on waste water handling.

2.4 Technical and professional organizations

- N. V. Vuilafvoer Maatschappij (VAM) (Waste Division state-owned waste division limited liability company)

  The capacity of the present company is around 1 000 000 t of mainly domestic waste and comparative industrial waste, from different municipalities. 180 000 t is separated mechanically, the rest is disposed of in sanitary landfill.

  The VAM company has been operating since 1929, with a long tradition and experience in composting. Nowadays, the VAM is planning to compost the organic fraction that is collected separately and to burn the rest. Apart from operational activities, VAM is experienced with waste handling.

  Director:
  Ir. W. de Feyter
  P.O. Box 5380
  1007 AJ AMSTERDAM
  Tel. 0 20 730307

- N.V. Afvalverwerking Rijnmond A.V.R. and N.A. A.V.R.-Chemie A.V.R., mainly owned by the Rotterdam township, incinerates (mainly domestic) waste, total capacity about 1 000 000 t/year. A.V.R. Chemie is an incineration plant for chemical waste, owned by eight industrial companies, the state and the Rotterdam township. It consists of two rotary-kilns with energy-recovery.

  Apart, A.V.R. Chemie runs and controls a C2-waste depony. The total process capacity of A.V.R. Chemie is around 160 000 t/year.

Director:
Ir. G. J. R. Nales
Prof. Gerbrandyweg 10
3197 KK ROTTERDAM-BOTLEK
Tel. 0 1819 42911

- Koninklijk Instituut van Ingenieurs—KIVI (Royal Institute for Engineers)
  Address:
  Prinsegracht 23
  2514 AP DEN HAAG
  Tel. 0 70 646800

  KIVI is in general, a consultant and technician association, whose members operate in all branches of science and technics under which environment and environmental protection.

- Vereniging van Nederlandse Gemeenten (VNG) (Association of Dutch local Authorities). The VNG board consists of local authorities. The VNG operates an office with one head director and three directors.
  Office address:
  De Willemshof
  Nassaulaan 12
  2514 JS DEN HAAG
  Tel. 0 70 738393

2.5 Addresses for sanitary landfills

- **NVAM** (see before under addresses on waste treatment in general).
- **NV RAZOB**
  Processing around 600 000 t/year, mainly domestic and comparative industrial waste.
  Director:
  Ir. L. H. J. van Beers
  Gulberg 9
  5674 TE NUENEN
  Tel. 0 40 835889
- **S.W.F.**
  Processing around 175 000 t/year, mainly domestic and comparative industrial waste.

Director:
Ing. D. Louwman
P.O. Box 201
1620 AE HOORN
Tel. 0 2290 41444

- **Tilburg**
Processing around 500 000 t/year, mainly domestic and comparative industrial waste.
Address:
Ing. J. Mertens
Ceramstraat 6
5013 BB TILBURG
Tel. 0 13 360000

- Vereniging Nederlandse Particuliere Reiningsbedrijven (Association of Dutch private waste management organizations) (VNPR)
Secretary General:
Mr. J. C. van Zundert
P.O. Box 90154
5000 LG TILBURG
Tel. 0 13 654200

President:
L. G. M. van Gansewinkel
Director-General: Gansewinkel BV
Den Engelsman 4
MAARHEEZE
Tel. 0 4959 2277

Members of the VNPR are private entrepreneurs working in the waste management and public cleansing field who submit themselves to a code of ethics.

- Vereniging van Exploitanten van Afvalverwerkingsbedrijven in Nederland (Union of Operators of Waste Processing Installations in The Netherlands).
Secretary General:
Mr. A. L. Cardon
Coolsingel 40
3011 AD ROTTERDAM
Tel. 0 10 4173459

The VEABRIN consists of a general board composed of administration officials and an advisory committee, composed of technical managers of the (11) installations.

Chairman of the advisory committee:
Drs. D. T. Folmer—director GEVUDO
P.O. Box 550
3300 AN DORDRECHT
Tel. 078 142433

- Nederlands Instituut van Direkteuren en Ingenieurs van Gemeentewerken (NIDIG) (Netherlands Institute of Directors and Engineers of Public Works)
  President:
  Ir. L. van Unen
  Director Public Works and Public Cleansing
  Stadhuisplein 2
  4382 LG VLISSINGEN
  Tel. 0 1184 67342

  The members of NIDIG are mainly managers of Public Works, generally on municipality level.

3.0 Spreading of Information

3.1 Scientific and technical journals

- Gementereiniging en afvalmanagement (GRAM) (Public Cleansing and Waste Management)
  Edited by M.S.G. BV in cooperation with NVRD (see Section 2.1)
  Address of editorial staff:
  Ing. J. A. Mertens—Chief Editor
  P.O. Box 649
  5000 AP TILBURG
  Address for Advertisements and Subscriptions:
  M.S.G. b.v.
  Julianaplein 21
  5211 BB 's HERTOGEN-BOSCH
  11 issues per year for about 500 pages, first published in 1907.

3.2 Exhibitions

- ENVIRO, organized every two years in RAI-Congresscentre Amsterdam by ENVIRO in cooperation with NVRD and some other organizations.

Address:
ENVIRO
Europlein 8
1078 GZ AMSTERDAM

The exhibition comprises environmental protection and waste management. There will be held, with ISWA sponsoring, special seminars in the field of solid wastes management and public cleansing.

- AQUATECH organized every two years in RAI-Congresscentre Amsterdam by AQUATECH in cooperation with organization working in the field of water technology.
  Address:
  AQUATECH
  Europlein 8
  1087 GZ AMSTERDAM

The exhibition comprises all aspects of water technology.

Table 1 Waste production in the Netherlands

See section	Categories	Quantities produced[a] ($\times 10^3$ t)
4.1	Domestic waste	4100
4.2	Bulky domestic waste	700
	Litter, market refuse, garden refuse and floating refuse	1000
	Mud, including mud from excavation	400
	Waste from production and processing	2000
	Waste from shops, offices and services	1500
4.5	Scrapped motor vehicles	500
	Car tyres	50
4.6	Construction and demolition waste	7000
4.7	Sewage sludge	5000
	"Agriculture waste"	500
4.9	Hospital waste	150
4.10	Waste gypsum	2000
4.11	Dredging spoil	50 000–60 000[b]
4.12	Combustion residues from coal-fired plants	570
4.13	Combustion residues from refuse incineration plants	700

[a] Quantities produced. Source: National Institute of Public Health and Environmental Protection/Waste Substances and Emissions Laboratory (RIVM/LAE).

[b] Total quantity of spoil dredged, of which 15–20 million m^3/year must currently be stored on land.

4.0 Production and Composition of Solid Wastes

Waste production in the Netherlands is shown in Table 1.

The Waste Substances Act and the Chemical Act have the following basic rules:

- The municipality is responsible for the domestic waste collection, at least once a week at every residence.
- The region (a "natural" combination of municipalities) is responsible for the waste disposal.
- The waste producer is responsible for its own waste.

4.1 Domestic waste

The composition of domestic waste is shown in Table 2.

Table 2 Average composition of domestic waste

Material	(%)
Paper, cardboard	24.0
Glass	6.7
Iron (mainly tinplate)	2.2
Non-ferrous metals	0.6
Plastics	6.5
Wood	0.7
Fruit, vegetables, garden waste	50.3
Bread	1.6
Animal waste	1.3
Textiles, rope	2.2
Carpets, mats	0.3
Leather, rubber, linoleum	1.1
Masonry, pottery	2.1
Other waste	0.4

4.2 Bulky domestic waste

Bulky domestic waste, originating from individual households, is too large or too heavy to be offered to the refuse collection service in the same way as ordinary domestic waste. The approximate composition of bulky domestic waste in percentage weight is:

27% paper
18% metal
2% plastics and rubber
23% wood
26% textiles
4% various inert materials

4.3 Beverage packaging

The proportion of domestic waste represented by packages gradually rose during the seventies to approx. 27% in 1979, since when it has fallen. A considerable proportion of domestic packaging waste consists of beverage packages, and it is this waste that has fallen considerably in recent years owing to the collection of waste glass in bottle banks. The decrease in the amount of packaging waste in domestic waste to 21.9% in 1982 is therefore attributable mainly to the increasing collection of waste glass and also to the introduction of lighter packages (savings in materials such as tinplate and glass, replacement of glass by paper/cardboard and plastics). By far the greatest proportion is related to soft drinks and beer, 95% and 90% respectively as far as containers for retail sale are concerned.

The number of returnable glass bottles used for dairy products has fallen steadily and is now approx. 30%. Returnable bottles for other types of drink hardly exist nowadays. Wine and spirits are sold exclusively in disposable bottles, although small-scale experiments are taking place with the return of bottles for reuse and the introduction of deposit bottles in the wine sector.

4.4 Stray waste

One facet of the environmental problem is the pollution of the environment by waste substances. In streets and squares, parks and gardens, woods, along verges and banks, ditches and canals and recreation grounds, one is confronted with this form of environmental pollution.

The quantity of stray waste that has to be disposed of is estimated at 330 000 t/year. This waste arises through human activities or through the absence of the appropriate activities, e.g. through the supply of transportation of waste but also through storage, processing and clandestine dumping, or usually through thoughtlessness in everyday activities.

Approximately 200 000 t of stray waste are collected by municipal sanitation departments as litter and market waste, 50 000 t are removed from waterways and parks, 20 000 t are collected by public bodies such as the State Forest Service, Dutch Railways, recreational enterprises and the like, and finally 60 000 t of stray waste are gathered while cleaning up the countryside.

4.5 Scrapped motor vehicles

In 1978, 360 000 scrapped motor vehicles were stored in 1656 scrapyards with a total area of 5 800 000 m^2. In order that scrapped motor vehicles may be disposed of in an environmentally-responsible way, the provincial authorities have a duty under Article 21 of the Waste Substances Act to draw up a provincial plan for the disposal of scrapped motor vehicles.

Most scrapped motor vehicles in the Netherlands are processed by a limited number of shredder plants. In the processing of scrapped motor vehicles, the mean weight of which may be said to be 800 kg, 70% can be recovered as iron and steel and 10% can be recovered as non-ferrous metals. The remaining 20% must still be dumped.

4.6 Construction and demolition waste

The quantities of construction waste that arise during construction projects show wide variations. Research has shown that approximately 1 million m^3 of construction waste are generated annually, of which about half comes from the housing sector. The amount of demolition waste generated annually is approximately 6 million t. About 80% of this demolition waste consists of concrete and masonry rubble; the remaining 20% consists of wood, metals, plastic, etc.

The reuse of asphalt has increased sharply in recent years, not least because of the favourable financial aspects. Technical developments have reached the stage where a large percentage of the broken asphalt can be reused.

The quantity of concrete and masonry waste suitable for reuse can still increase considerably (to at least 4 million t/year). Before this can be achieved, however, either selective demolition methods must be used on a much larger scale or the processing plants must be equipped to separate not only metals but also wood, plastics and other light material from the rubble. A further and particularly important requirement is that the market will accept quantity of broken material.

4.7 Sewage sludge

There are about 500 sewage purification plants in operation at which sewage is purified mechanically and/or biologically. The total quantity of sewage sludge produced by the plants is around 6 million m^3 with a mean dry-matter content of 4.4% (= 255 000 t dry matter).

Additional sludge is produced by purification plants that form part of industrial establishment. A considerable proportion of the sewage sludge is dumped. For dumping of the sludge to be acceptable, the dry-matter content must be such that the dump mass is impenetrable and no water can seep from it during or after dumping.

Sewage sludge is also used to provide interim or final covering layers for dumps. Due to quality requirements of the environment, the incineration of sewage sludge is applicated at a growing scale.

4.8 Specific organic waste substances

This category of waste amounts to approximately 500 000 t excluding animal manure. Part of this waste can be used as cattle fodder, but a certain proportion including waste with a high moisture content from asparagus, cucumbers, leeks, ornamental plants and pot plants is not suitable for this purpose and is dumped.

There is a high production of animal manure, due to the intensive cattle and poultry breeding.

There are (1985): around 5 000 000 cows
around 11 000 000 pigs
around 84 000 000 chickens

with a total production of around 85 000 000 t of manure. Some tight measures are taken by the government to diminish the quantities of manure.

4.9 Hospital waste

This category is understood to include all waste substances that are generated in establishments for intramural health care insofar as such waste substances do not fall within the scope of other acts. In addition, certain industrial waste substances from establishments engaged in the production of sera and vaccines and in pharmaceutical and toxicological research, etc. are of a similar character and composition to hospital waste.

The total quantity is currently estimated at 150 000 t, whereas the estimates in 1972 gave a total of 75 000 t/year. There are differences among different types of establishments. In special hospitals and university hospitals, the quantity of waste per bed is considerably greater than in psychiatric hospitals and nursing homes (6 kg as opposed to $10\frac{1}{2}$ kg/day).

4.10 Waste gypsum

Gypsum is generated in the chemical industry as a by-product of the manufacture of products such as phosphoric acid, hydrofluoric acid, magnesium oxide, citric acid, boric acid and titanium dioxide. In addition, there is also the flue-gas-desulphurization-gypsum that is obtained (via calcium sulphite) in the desulphurization of the flue gases of coal-fired boilers.

The types of chemical gypsum generated in significant quantities in the Netherlands are phosphoric-acid-gypsum (phosphogypsum), magnesia-gypsum and flue-gas-desulphurization gypsum, which currently amount to about 2 million t/year.

In view of the quantities generated and the associated problems, attention has been almost entirely directed towards the processing of phosphogypsum. Most of this is currently still discharged into surface water (Waterweg, Westerschelde), but these discharges will have to be abandoned within the foreseeable future, particularly in view of the presence of cadmium. Attempts are therefore being made in the first instance, to find possible applications for phosphogypsum. The possibilities and preconditions for the storage of phosphogypsum on land are also being investigated.

The research programme of the working group on phosphogypsum currently consists of nine projects; parts of which are already complete. The following subjects may be distinguished:

- A mixture of phosphogypsum and fly ash with cement as a binder may perhaps find application in stabilization layers in road building. Some trial surfaces have been laid, and the result has led to the preparation of a new project on a somewhat larger scale.
- A granulate made from phosphogypsum, fly ash and cement may perhaps be useful in the construction of embankments and for raising the land level.
- If gypsum products can be made water-resistant, there will be a much wider range of possible applications.
- A significant development is the possibility that alpha-hemihydrate may be produced from phosphogypsum. This type of gypsum has a particularly high strength and a very short setting time.

In addition, research is being pursued into a new phosphoric acid preparation method in which heavy metals are removed. This process, which is being tested on a laboratory scale, has potential positive consequences for the composition of the residual phosphogypsum.

4.11 Dredging spoil

Silt occurs as a natural component in all Dutch waters. The extent to which it occurs, its composition and its content of pollutants vary from area to area; differences occur as a result of seasonable phenomena and discharges. Where a river becomes wider or deeper, the water flows more slowly and more of the silt and sand particulates entrained by the water will settle. The material obtained when waters are dredged is usually called dredging mud or dredging spoil.

It is estimated that about 40–50 million m^3 dredging spoil is removed from national waters annually in dredging activities in rivers and access channels, particularly in channels that provide access to seaports. The locations where the largest annual quantities of dredging spoil are produced are:

- The Westerschelde (15–20 million m^3); much of the spoil is dumped back into the river elsewhere;
- The IJmond at IJmuiden (3–5 million m^3); virtually the entire quantity of spoil dredged from the sea is dumped back into the sea;
- The Ems estuary (approx. two million m^3); this spoil is dumped in the Bocht van Wattum and in the mouth of the Dollard;
- The Rhine estuary (20–25 million m^3); the mud that is dredged in this region has a very variable degree of pollution and is classified into the following quality classes in increasing order of pollution:
  - Class 1: 14 million m^3; spoil from the western harbour area (Europort, Maasmond and the Nieuwe Waterweg);
  - Class 2: 4.2 million m^3; spoil from the middle harbour area;
  - Class 3: 3–4 million m^3; spoil from the eastern ports and rivers;
  - Class 4: approx. 0.5 million m^3; locally seriously polluted spoil from the eastern and middle harbour area.

A policy plan for the disposal of dredging spoil has been drawn up which shows the affected authorities, the policy to be followed in the short-, medium- and long-term. The policy in the short-term is to find a limited number of relatively large storage locations. A suitable site was recently found on the Maasvlakte: approx. 18 million m^3 of dredging spoil can temporarily be stored there in the short-term. In the medium-term, an attempt is being made to find a large-scale coastal location, while in the long-term, the primary emphasis is being placed on the possibility of preventing the contamination of the dredging spoil by aiming for a sharp reduction in the discharges of substances.

4.12 Combustion residues from coal-fired installations

The three coal-fired power stations produce around 500 000 t of fly ash. These waste substances are characterized by their quantity, the possibility of producing a constant quality and the possibility of centralized processing at the coal-fired installation. The electricity producers who operate coal-fired power stations have set up a cooperative association, Vliegasunie BV, for the marketing of fly ash.

In addition to this quantity of waste substances, a smaller quantity is generated which consists of fluidized-bed ash and residues from industrial travelling-grate and pulverized-coal installations. This ash is of variable quantity and the possibilities for processing it directly in the coal-fired installation are more limited. This ash amounts to approx. 90 000 t.

Research is in progress into new firing and conversion techniques such as fluidized-bed combustion and coal gasification.

4.13 Combustion residues from waste incineration plants

After incineration, about 30% of the mass of the original waste remains, consisting of slag and fly ash in a ratio of 7 : 1. The slag and to a greater extent the fly ash, are susceptible to leaching and can contain elements such as lead, cadmium and zinc in concentrations that exceed the limits stipulated in the Chemical Waste Act. From research carried out at the Waste Substances Research Institute, it appears that the heavy metal molybdenum is leached out to some extent (1–10%) from slag and fly ash and that cadmium and chlorine are leached out to a considerable extent (10–90%) from fly ash. In the Substances and Process Decree under the Chemical Waste Act, combustion residues from facilities for the incineration of domestic waste are not regarded as chemical waste.

Research into the composition and leaching behaviour of slags and fly ash has also been carried out by the Technical Advisory Bureau of Association of Water Control Boards (1983) under contract with the management boards of the waste incineration plants.

The main incentive for this study was the stagnation of the market in slags, which therefore often found application in road-building, etc. This application has diminished as a result of a more stringent soil protection policy.

Fly ash is used in a bonded form in asphalt and concrete. Recycling of fly ash as a filler in asphalt mixtures has become both technically possible and environmentally acceptable. A filler of this type has been certified by the Institute for the Control of Fillers. Part of the fly ash that is generated is still disposed of by controlled dumping.

5.0 Collection of Domestic Waste

The organization of the collection is shown in Table 3.

Due to a growing political demand for privatization, the collecting by private entrepreneurs will be higher today than in 1979. Collection is done by plastic bags (30%) and by minicontainers (70%). Glass is in general collected separately by means of glass boxes. Around 35% of the glass input is recycled in this way. Due to increasing application of waste incineration, there is a growing tendency to collect separately in two flows the "dry" waste components and the "wet" components. Several experiments are on, with promising results.

Table 3 Organization of collection service (1979)

Service	% Municipalities	% Inhabitants
Municipal collection service	38	74
Collection by neighbouring municipality	10	3
Regional collection service	16	6
Collection by private enterprise	36	17

6.0 Waste Treatment and Disposal

6.1 Incineration

In the Netherlands, there are 11 government-controlled incineration plants for the destruction of domestic waste and other waste (2.4 million t/year). A number of these make a contribution to the energy supply. Incineration can also make a modest contribution to the conservation of raw material, e.g. in road-building.

Incineration in a grate has proved to be the most practicable method of incinerating domestic waste. In addition, there are 51 installations with a total of 59 furnaces for the incineration of waste substances that are generated on their own premises. For 48 of these furnaces, the total capacity is 92 t waste/h. The capacity of the remainder is unknown. Most of these furnaces are located in the timber and furniture industry. In more than half of them, the heat that is released is exploited. These installations are omitted from the following discussion, partly because of lack of data.

Fluidized-bed combustion has the advantage of better heat transfer and material transfer. Owing to the improved heat transfer, the system is insensitive to variations in the composition of the fuel and the heat is used more efficiently. The improved transfer permits harmful emissions, particularly SO_2 emissions, to be lowered by appropriate addition of chemicals; very little is known about the limitation of microemissions by this technique, however.

The Central Organization for Applied Scientific Research in the Netherlands (TNO) and the Netherlands Energy Research Foundation (ECN) are currently carrying out emission studies on the incineration of selected fractions of domestic and industrial waste. Large-scale incineration of domestic waste in fluidized-bed incineration plants has not yet been introduced in the Netherlands.

Gasification and degasification processes can be described as the thermal breakdown of organic substances in the presence of little or no oxygen. The heat required for degasification is added by means of a hot inert medium (sand, gas) or by external heat. In gasification, a portion of the heat that is necessary for the process, is generated internally by combustion of part of the organic material in the reactor (partial oxidation). These liquids generate gases, liquids and carbon-containing residues, which can be exploited as stable, transportable energy carriers.

Subsidized research in the field of thermal processing of waste substances is being carried out on the cocurrent gasifier at Twente University of Technology and on fluidized-bed pyrolysis at Eindhoven University of Technology. The purpose is to compare these methods with regard to a number of aspects such as environmental effect, raw materials and costs.

6.2 Sanitary landfill

45% of all domestic wastes are disposed of in sanitary landfills.

The situation is as follows:

Number of landfill with a yearly capacity

From	5000– 50 000 t:	29
	50 000–200 000 t:	54
	200 000–500 000 t:	8
	4 500 000 t:	2

There are official guidelines and regulations to design and to manage controlled landfill sites with high standards.

- Upper- and/or underliners are necessary in places where natural clay is missing. Leachate control is obligatory. Geophysical detection methods are used to compare with direct ground water measurements.
- There is a tendency for closing the smaller landfills and concentrate on the bigger ones. Gas production from landfills is increasingly applied.
- Land restoration or reclamation to beneficial after-uses (public amenity or agricultural purposes) has to be foreseen in an acceptable environmental impact in a long perspective of time.
- Health and safety aspects concerning landfill workers and development of landfilling techniques and equipment are planned in advance and evaluated afterwards.
- The problems of "yesterday's wastes" has also a big attention in the Netherlands because of the environmental impacts. Ground water problems have to be solved. Special techniques are developed (upperliners, bentonite screens and so on).

Table 4 Incineration plants operated by the central government (1982)

						Recycled						
						Iron				Electricity		Distilled water produc- sold, 1000 m^3
	Capacity of plant	Con- structed in	Inhabit- ants in catchment area,	Incinerated		Generated		Sold		Generated	Sold	
Plan/region	(t/h)	(year)	1-1-81	tons	kg/inhab.	tons	%	tons	%	MWh	MWh	
Alkmaar	3 furnaces of 6	1971/ 1978	265 436	111 250	419	3790	3.4	3790	100			
Amsterdam	4 × 16	1968	897 518	386 959	430	9421	2.4	9421	100	149 534	137 506	
Arnhem	3 × 12	1975	364 626	165 804	455	3855	2.3	3855	100		—	
Dordrecht	3 × 7	1972	362 100	110 359	305	1931	1.7	1931	100		—	
The Hague	4 × 12.5	1967/ 1974	581 070	280 532	483	7787	2.8	7787	100	90 404	76 077	
Leeuwarden	2 × 6	1973	133 386	63 693	478	1941	3.0	1914	100			
Leiden	3 × 4	1966/ 1976	227 572	77 500	341	slag not segregated						
Roosendaal & Nispen	2 × 4	1976	88 021	32 894	374	slag not segreated						
Rotterdam	4 × 13	1963	568 200	259 100	456	8177	3.2	8177	100	86 317	77 383	
A.V. Rijnmond	6 × 20	1972	1 573 604	777 815	494	20 000	2.6	18 817	94	183 805	124 724	6044
Zaanstad	2 × 9	1976	257 077	108 356	421	4662	4.3	4662	100			
	Total		5 318 610	2 373 262	446	61 537		60 364		509 060	415 690	6044

11

Norway

Surface:	323 895 km^2 excluding Svalbard
Population:	4 180 000 inhabitants
Counties:	19
Municipalities:	450

1.0 Waste Policy

Until recently, the environmental authorities have concentrated their efforts mainly on the clean-up and repair work connected with the damages already done to the environment. Along with the continued repair of existing damages done to the natural resource base, it has now also been possible to give more attention to preventive environmental protection. A central task in this respect will be to strengthen the cooperation between local and national authorities in matters regarding the environmental policy. Important elements will be the establishing of national resource managements and effective measures to reduce pollution. The environmental and resource policy will now get a more long-term profile, through considerable increases in allowances for research and special long-term studies. More funds will also be provided for management activities in high priority areas.

Environmental considerations will form an important part of all future sector planning in Norway. Integration of various environmental concerns in the sector planning and decision-making activities will therefore be of primary importance.

As part of the new environmental policy, a status report on the environment containing information regarding the utilization of natural resources will be issued annually. More stringent standards for emissions from industry, agriculture and municipal waste and sewage treatment will also be brought into effect.

In 1984, the Ministry of Environment presented an action plan for recycling. This plan is based on the idea that a number of environmental problems can be solved through recycling, thus increasing the resource base for our society. This is based on the following three major principles:

- The amount of waste should be limited by careful planning at an early stage of the production process, through internal recycling of materials, increased life-span of the products, increased maintenance, effective return and recycling systems and by establishing a market for second-hand goods.

- Raw materials based on waste-materials must have guaranteed and stable market conditions, and if possible used in preference to new raw materials.
- Recycling should be carried out at the highest possible level of refining in relation to the original raw material.

The action plan is based on the Pollution Control Act which states that waste should be recycled whenever a combination of environmental considerations, resource conservation and economical factors makes it feasible.

The long-term policy is that recycling of materials should be incorporated in product-design and processes. Recycling should also be considered a normal method of waste treatment. A number of new measures are necessary, however, in order to stimulate increased recycling at present.

In order to meet these objectives, the plan contains an important decision about the possibility to obtain direct subsidies to establish pilot plants and pilot projects. The annual amount is approx. 2.5–3.0 million NOK.

In the first phase of the plan, four different types of raw materials have received priority: organic waste, waste paper, plastic waste and car tyres. From 1986, all waste materials will be treated equally. However, in the initial phase, most projects will still pay most attention to the four materials mentioned.

In the four priority areas, several projects have already been established. Research has been going on for some time already concerning the use of fish wastes and food processing waste as animal feed through recycling. Development projects are also carried out in the field of collection and storing of such wastes in order to make the handling of raw material more efficient and hygienic.

Recycling of waste paper aiming to produce lower grade paper has already been in operation for several years. Projects are now established regarding the use of waste paper in new markets such as in the product of insulation materials from old newspapers. Such products are already well established on markets for instance in the USA and Canada.

Four companies using plastic wastes as raw materials are established. Three are converting plastic foils into a grandulate retailing it to manufacturers of other plastic products. One company has specialized in processing PVC-waste. This is an international novelty.

A most recent survey of waste handling in municipalities shows annual investment needs of 375 million NOK/year ($50 million). The amounts will vary according to the methods being used. At the moment both recycling and incineration is being evaluated in several trial plants, while the use of the old traditional landfill method still continues. The ministry, however, has estimated that approx. 60% of all local waste dumps should have been closed.

Concerning municipal waste disposal, the national authorities has put special emphasis on the importance of establishing regional cooperation. This also is an attempt to replace many of the small local facilities with modern well-equipped waste treatment plants.

2.0 Addresses

The Norwegian Association of Public Cleansing and Solid Waste (NRF) (National ISWA member).
Secretariat:
Oslo renholdsverk
Gunnar Værnes, general chairman
Gunnar Schjelderupsvei 11
0485 OSLO 4
Tel. 02 22 71 50

The Ministry of Environment
Pollution Control Department
Box 8013, Dep. 0030, OSLO 1.
Tel. 02 11 90 90

State Pollution Control Authority (SFT)
Attn. Mr. O. M. Grini
Box 8100, Dep. 0032, OSLO 1
Tel. 02 65 98 10

Royal Norwegian Council for Scientific and Industrial Research (NTNF)
Sognsveien 72
0855 OSLO 8.
Tel. 02 23 76 85

Technical University of Norway
Institute of Hydraulics and Sanitary Engineering
Alfred Getz vei 3
7034 TRONDHEIM—NTH
Tel. 07 54 30 00

Norwegian Institute for Water Research (NIVA)
Brekkeveien 19
0883 OSLO 8.
Tel. 02 23 52 80

Norwegian Institute for Air Research (NILU)
Elvegaten 52
Postbox 64
2001 LILLESTRØM
Tel. 06 81 41 70

The Foundation of Scientific and Industrial Research (SINTEF)
7034 TRONDHEIM NTH
Tel. 07 59 30 00

3.0 Quantity and Composition of Solid Waste

The Pollution Control Act defines three waste categories: consumer waste, production waste and special waste. Consumer waste is waste generated in households, offices, shops, institutions and similar. This corresponds closely to the term municipal waste, which is waste normally handled by the ordinary waste collection and treatment system. Production waste is waste from manufacturing and industrial processes. Special waste is waste needing special handling because of health hazard/environmental risks (hazardous waste) or because of size of physical properties (bulky waste). Normally, this term is synonymous with hazardous waste.

The total amount of consumer waste (municipal waste) is 1.9 million tonnes annually, from which approx. 0.8 Mt is generated in households. The composition of the municipal waste is shown in Table 1.

Table 1 Composition of municipal waste

Category	Percentage of total (%)
Paper and cardboard	31
Food waste and other putrescibles (organics)	18
Plastics, textiles, rubber	7
Other combustibles	21
Glass and metals	8
Other noncombustibles	15

The household waste in Norway is characterized by a high percentage of paper and putrescibles. The average composition is given in Table 2. Considerable regional and local variation from this average can occur.

The low metal- and glass content in the household waste is mainly due to the widespread use of returnable bottles for soft-drinks and beer.

There are considerable variations within the country regarding the composition and the generation of municipal waste. These variations are evident from Table 3, which presents the variations within the waste-categories included in the municipal waste.

Data on production waste handled by industry itself is scarce. It is estimated however to be approximately equal to the total amount of municipal waste, i.e. around 2 million t/year. This does not include mining waste or waste from various earthworks.

Table 2 Average composition of household waste

Category	Percentage of total (%)
Paper and cardboard	33.5
Food waste (putrescibles)	30.5
Textiles	3.1
Rubber and leather	1.1
Plastics	5.1
Other combustibles	5.3
Metals	4.0
Glass	5.1
Other non-combustibles	1.4
Fines (<10 mm)	2.9

Table 3 The composition variations of municipal waste

Category	Percentage of total (%)	kg/inhabitant/year
Household waste	35–45	90–250
Bulky waste from households	1–3	10–20
Waste from shops and offices	5–15	25–80
Light industrial waste	15–25	10–150
Construction and demolition waste	10–20	70–100
Other waste	8–13	15–100
Total	—	220–700

The amount of special waste (hazardous waste) is estimated to be 120 000 t annually. A comprehensive survey of the generation of hazardous waste will soon be available. Of the total amount of 120 000 t hazardous waste about 100 000 t is presumed to be different types of oil waste, another 15 000 t consists of organic compounds partly chlorinated, while hazardous inorganic wastes only amount to approx. 5000 t.

4.0 Waste Collection

According to the Pollution Control Act, the collection of consumer waste is a municipal responsibility. About 60% of consumer waste is collected by private firms on special contracts with the municipal agency. However the municipality itself is always responsible for collecting the fees. Production waste is mainly collected by private transportation firms or by the industry itself.

Approximately 90% of the population is covered by a regular waste collection service. In central areas with more than 2000 inhabitants, 100% of the population is covered by such service. The normal collection frequency is once a week. It is not uncommon however to have 3 collections weekly in the cities. The use of bags for the collection is still most common. During the last few years however, the use of plastic bins on wheels has increased considerably. Modern rear-loaded-high-compaction vehicles are commonly used for the refuse collection service. Normally, the collection truck delivers the waste directly to the waste disposal plant. The use of transfer stations or satellite vehicles is not yet common in Norway.

Source separation of household waste has been practised in two cities so far, as part of a test programme. Separate collection of newspapers and magazines however has been practised to some extent in the most populated areas.

Separate collection of waste paper, especially cardboard, is fairly common from warehouses, offices and some industry. There also exists a rather extensive use of separated food-waste from restaurants, institutions and similar. The use of source separation in general will probably increase in the near future.

5.0 Treatment and Disposal of Municipal Waste

The most common disposal method up-to-date has been landfilling. The typical pattern has been one separate landfill in each municipality. During the last 10–15 years, several new regional disposal facilities have been established. Some of these are based on incineration, composting or modern landfill methods. As a result of this, many small disposal sites with severe negative environmental effects, have been closed.

At present, there are about 380 municipal waste facilities in Norway. The distribution of the various waste treatment methods is presented in Table 4.

Most of the incineration plants are small and without energy recovery. These were built during the "incineration wave" in the 1960s. The incineration method has attracted a new interest in the last few years and five new large and modern incineration plants are now in operation. All of these are designed for energy recovery in the form of district heating, in some cases, combined with the production of electricity. It is expected that incineration will play a more dominant role in the total waste treatment system in the coming years.

Table 4 The use of various waste treatment methods

Type of facility	Number	Percentage of waste treated (%)
Landfills for MSW	265	53
Landfills for bulky waste	50	12
Incineration plants	50	23
Milling plants	10	8
Composting plants	5	4
Total	380	100

Most of the milling plants were built in the early 1970s when the method was very popular. Some of these plants have already been shut down. A few have been upgraded to composting plants. Some of these are fairly new. Composting, however, is not considered as a suitable waste treatment method in Norway and no new plants are planned.

One of the original milling plants has been redesigned to produce RDF-briquettes based on the Swedish BRINI-system. The plant processes 30 000 t waste annually producing 13 000 t briquettes. The RDF is utilized in a special fluidized-bed boiler at a nearby paper mill. The burners, used for drying the briquettes, are fuelled by biogas generated at a nearby landfill.

In the city of Oslo, a central sorting plant was opened in 1985. The total investment was approximately Norwegian kr 550 million. The plant was designed to process 140 000 t of mixed household waste annually by mechanical sorting. The sorted materials are paper (32 000 t/year), plastics 3500 t/year) and metals (5000 t/year). The remaining fraction is burned together with 40 000 t commercial waste at an incineration plant integrated in the sorting plant. Up to this date, the output of products and the quality has been substantially lower than scheduled.

This sorting plant in Oslo is the only one based on mixed household waste. There are, however, a few other plants processing commercial waste. These are mainly based on manual sorting and the main product is cardboard and to some extent, metals and wood. The market is showing a growing interest for sorted commercial and industrial waste.

During the recent years, rather strict regulations have been made on waste disposal in general. For landfills, standard requirements include sealing to the underground and a proper drainage system for leachate. Normally, leachate must be treated before discharge either in a separate treatment plant, or combined with municipal waste water.

Up to 1986, restrictions on emission from waste incineration were rather "mild". The only regulation was on dust emission, where the limit was maximum 100 mg/nm^3 for large incinerators and 250 mg/nm^3 for smaller plants. From July 1, 1989, a new set of requirements will be effective. These are presented in Table 5.

Table 5 New requirements to incineration plants

Parameter	Emission limit (mg/nm^3)
Dust	30
Hydrogen chloride (HCl)	100
Sulfur dioxide (SO_2)	300
Mercury (Hg)	0.1
Cadmium (Cd)	0.1
Carbon monoxide (CO)	100
Dioxines (only guidelines)	2 mg/nm^3 TCDD-equivalent

In order to meet these new emission standards, all existing large incinerators must be equipped with an advanced gas cleaning system. In addition to the requirements in Table 5, the new standard also gives strict regulations to the landfill areas being used for the deposition of fly ashes from the gas cleaning system.

To reduce the amount of particularly cadmium and mercury in the waste, in general, the authorities are pushing the municipalities to start separate collection of used-batteries. In 1986, such collections started in two different municipalities in Norway.

6.0 Industrial and Hazardous Waste

Most of the industrial waste not treated at municipal facilities, is burnt for energy recovery, reclaimed or deposited at special private landfills. In all wood-based industries, which is of special importance in Norway, burning with heat recovery is extensively used as a method to handle various waste products as bark, saw-dust and black liquor.

Another example of utilization, is the use of dust from metallurgical processes in the cement industry.

The waste from food-processing industry is to a large extent converted into animal feed.

The mining industry has in some locations, severe difficulties dealing with environmental problems from deposition of tailing and other waste materials.

Hazardous waste is treated by different methods in various industrial plants. There is no central waste treatment facility in Norway similar to those established in the other Scandinavian countries. Partly as a result of this, certain types of wastes therefore must be exported in order to obtain an adequate treatment.

Most of the hazardous waste consists of a type of combustible oil. A few industrial boilers are authorized to burn this type of waste. Also, a rotary kiln in the cement industry is being used for burning liquid waste. This method has proved to be very effective in the destruction of various chlorinated organics. A special plant that emulgates solid or semi-solid organic wastes has been established. The emulgation is later used as fuel in authorized boilers. Until recently a refining plant for used motor oil was in operation. The plant processed 6–8000 t/year of waste oil, but had to shut down for economical reasons.

Metal hydroxide waste and certain other types of inorganic waste should be deposited at one of six special landfills.

Authorization from the State Pollution Control Agency is required in order to receive and process hazardous waste. There also exist certain regulations concerning collection and transportation. In this case, authorization is given by the local agency. The country is divided (from 1986) into seven hazardous waste managements (regions), each with one responsible agency.

A system of regional and local receiving and transfer stations for hazardous waste will be established during the next few years. At present, only a few such regional waste managements are in operation.

7.0 Legislation and Management

The central legislation in this field is the Pollution Control Act which came into effect on October 1, 1983. The Act is applicable to pollution of water, air and soil. It applies to the emission of gases, solids or fluids, to noise or vibrations and generally to any emission that might cause damage or have negative effects on the environment. Waste handling and treatment is regulated by this act.

The Act is based on the basic principle that pollution of the environment is forbidden, unless a specific permit has been granted or the activity is in compliance with general regulations issued under the Act. The pollution control authorities will coordinate their activities with the land-use planning authorities in order to make the legislation pertinent to limit the pollution and waste problems effective.

The Act also provides for actions regarding waste recycling and other waste treatment methods, as well as the establishment of municipal sewage treatment plants. It also regulates actions related to incidents of acute pollutions such as oil spills or accidents in the handling of chemicals. At present, regulations concerning various measures to combat large oil spills have been issued under the law. Similar regulations regarding accidents involving other chemicals have not yet been brought into effect.

The Act approves various sanctions including fines which can be put into effect if emission takes place in violation to given permit or standard. The Pollution Control Authority is authorized to set the amount of the fines so high that it will not be practical for the polluter to continue the violation.

According to the Pollution Control Act, the collection and disposal of consumer waste is a municipal responsibility. The municipalities can also issue necessary local regulations concerning the waste handling in order to ensure an effective and hygienic waste collection system.

The collection and disposal of hazardous waste is regulated by special directions issued by the State Pollution Control Authority. They state an obligation to deliver all hazardous waste to authorized disposal sites when the generation rate is above the specified level.

The government authority with the ultimate responsibility for the Pollution Control Act is the Ministry for Environment. The ministry has delegated this authority to central, regional and local levels. The Pollution Control Council decides in accordance with the Act in cases, where assessments are necessary on environmental matters. The main administrative central authority is the State Pollution Agency. This authority has been given a wide power of attorney through the Act. The State Pollution Agency therefore in practice, administers the Act when no further delegations have been issued. On the regional level, authority has been delegated to the County Governor concerning the following area:

> Releases of municipal sewage from agricultural activity and fish processing, waste deposition from households, normal industrial production and sludges from sewage treatment plants, dumping of substances or articles in the sea or in lakes.

At the local level the municipalities give permits for the direct, local release of domestic effluent from individual houses, issue regulations and decide on matters regarding household waste collection and treatment. They also have the authority to order clean-up of illegal waste dumps or to order this done at the violator's expense.

To organize the handling of hazardous waste, the responsibility has been delegated to seven regional authorities. They report to the State Pollution Agency.

12

Poland

Surface: 312 683 km^2
Population: 36 914 000 inhabitants
Municipalities: 812 of which:

- 1 with more than 1 million inhabitants
- 38 with 100 000–1 000 000 inhabitants
- 328 with 10 000–100 000 inhabitants
- 445 with less than 10 000 inhabitants.

1.0 Addresses

1.1 ISWA national member

Polish Association of Sanitary Engineers and Technicians,
Czackiego 3/5, 00–043 Warszawa; Tel. 26 28 94

National Representative:

- Hanna Piotrowska, at the above-mentioned address.

1.2 Government offices

- Ministerstwo Budownictwa, Gospodarki Przestrzennej i Komunalnej (Ministry of Construction and Environmental Management),
  Wspólna 2, 00–926 Warszawa
- Ministerstwo Ochrony Środowiska i Zasobów Naturalnych (Ministry of Protection of Environment and Natural Resources),
  Wawelska 52/54, 02–067 Warszawa
- Ministerstwo Gospodarki Materiałowej i Paliwowej (Ministry of Materials and Fuel Management),
  Wspólna 1/3, 00–529 Warswawa

The Government generally coordinates the planning and practical realization of waste management systems depending on the 49 voivodeships (provinces) in the national territory.

1.3 Technical and professional bodies

- Polish Association of Sanitary Engineers and Technicians, Main Section of Public Cleansing and Waste Management,
  Czackiego 3/5, 00–043 Warswawa
- Local sections of public cleansing and waste management.

It is essentially an association of consultants and technicians, whose members operate in the field of solid waste management.

1.4 Main research institutions

- Instytut Gospodarki Przestrzennej i Komunalnej (Institute of Environmental and Municipal Management),
  Krzywickiego 9, 02–078 Warszawa
- Instytut Ochrony Środowiska (Institute of Environmental Protection),
  Krucza 5/11, 00–548 Warszawa
- Politechnika Warszawska, Instytut Inżynierii Środowiska (Warsaw Technical University, Institute of Environmental Engineering),
  Nowowiejska 20, 00–653 Warszawa
- Politechnika Wrocławska, Instytut Inżynierii Ochrony Środowiska (Wroclaw Technical University, Institute of Enviromental Protection Engineering),
  pl. Grunwaldzki 9, 50–370 Wrocław
- Politechnika Śląska, Instytut Inżynierii Ochrony Środowiska (Technical University of Silesia, Institute of Environmental Protection Engineering),
  pl. Powstańców Warszawy 10/13, 44–100 Gliwice
- Ośrodek Badawczo-Rozwojowy Oczyszczania Miast OBROM (Research and Development Centre of Public Cleansing),
  Okopowa 79/85, 91–850 Łódź
- Centrum Badawczo-Rozwojowe Gospodarki Przestrzennej (Research and Development Centre of Environmental Management),
  Filtrowa 57, 02–056 Warszawa
- Zakład Badawczo-Doświadczalny Gospodarki Komunalnej (Research and Experimental Plant of Municipal Management),
  Markiefki 24, 40–213 Katowice

2.0 Spreading of Information

2.1 Scientific and technical journals

- Gaz, Woda i Technika Sanitarna (Gas, Water and Sanitary Techniques),
  00–952 Warszawa, skr. poczt. 24;
  12 issues/year, 15% dealing with solid wastes,
- Gospodarka Materiałowa (Materials Management),
  00–950 Warszawa, skr. poczt. 128;
  24 issues/year, 15% dealing with waste management,
- Ogrzewnictwo, Ciepłownictwo i Wentylacja (Heating and Ventilation),
  00–950 Warszawa, skr. poczt. 1004;
  12 issues/year, 10% dealing with solid wastes,
- Zeszyty Naukowe (Scientific Bulletins)—Technika Sanitarna w Ochronie Środowiska (Sanitary Techniques in Environmental Protection), edited by Polish Association of Sanitary Engineers and Technicians.
  Czackiego 3/5, 00–043 Warszawa.

2.2 Exhibitions

There is no specific exhibition in Poland dealing exclusively with solid wastes. The subject is usually included in more general events concerning the various aspects of environmental protection. The most important taking place regularly are:

- Polski Kongres Oczyszczania Miast (Polish National Congress of Public Cleansing) organized every four years in Szczecin.
- Międzynarodowe Targi Poznańskie (Poznań International Fair) organized annually in Poznań.

2.3 Conferences

Conferences take place regularly on the occasion of the above-mentioned Polish National Congress of Public Cleansing. These conferences last two or three days and are Polish-spoken. There are several conferences and symposiums that do not occur regularly. Detailed notices about these events are reported in "Gaz, Woda i Technika Sanitarna" and current information of PZITS.

2.4 University courses

Solid waste management is generally included in the course programme called Sanitary or Environmental Engineering existing in the Faculty of Sanitary and Hydraulic Engineering of the main Polish technical universities.

There are also postgraduate courses entirely dealing with the problem of solid and hazardous wastes organized by Warsaw Technical University, Institute of Environmental Engineering.

3.0 Municipal Solid Waste Disposal

99% of solid wastes in Poland are disposed by landfilling. Only a small number (10%) of existing landfills may be considered as controlled landfills. The major number of Polish landfills are partially controlled or even free landfills. No more than 1% of solid wastes is disposed by composting.

In spite of a not very good situation in municipal solid waste disposal, a slow trend in improvement is presently occurring. The method of composting is being developed. Some composting plants have been planned, designed and are soon going to be constructed. The most significant are composting plants in Warsaw and Katowice. Besides composting a wide programme of controlled landfilling is being realized. Most of newly-organized landfills are lined and well equipped.

13

Portugal

Surface: 91 872 km^2
Population: 9 965 000 inhabitants (estimated in 1980)
Municipalities: 305 from which:

- 1 with more than 1 million inhabitants
- 1 with 300 000 inhabitants
- 5 with 100 000 inhabitants

22% of the population in 1915 nucleus between 2000 inhabitants and 500 inhabitants;
35% of the population in 36 425 hamlets with less than 500 inhabitants.

1.0 Addresses

1.1 ISWA National member

- Portuguese Association on Water, Wastewater and Solid Wastes; R. Antero de Quental, 44—1 100 LISBOA
  Tel. 54 35 44

1.2 Government offices

- Ministério do Plano e da Administração do Território (Ministry of the Planning and of the Territory Management, Rua do Século, n° 51, 1 200 LISBOA
- Direcção Geral do Saneamento Básico (Basic Sanitary), Rua Antero de Quental, 44—1 100 LISBOA

1.3 Technical and Professional Bodies

- Associação Portuguesa para Estudos de Saneamento Básico (Portuguese Association for Basic Sanitation Studies), Rua Antero de Quental, 44—1 100 LISBOA

It is a scientific association whose members are municipalities, municipal services, experts and individuals concerned on pollution and environmental matters. One of its branches is the Comissão Portuguesa de Lixos, a special Committee dealing with solid and hazardous wastes.

1.4 Legislation

According to the Decreto-Lei n° 488/85 (Decree 488/85), the jurisdiction on urban solid wastes (RSU—Resíduos Sólidos Urbanos) is enforced by the following bodies:

- Ministério do Plano e da Administração do Território:
  - To determine the national policies on RSU
  - To outline national and regional plans, general directives for transportation and removal, treatment and disposal of solid wastes
  - To determine the regulations for construction, installation and infrastructures for removal and treatment of solid wastes.

Within the Ministério do Plano e Administração do Território, the departments dealing with these matters will be

- Direcção Geral do Saneamento Básico
- Direcção Geral da Qualidade do Ambiente

Municipalities either as individuals or as associated bodies:

- To determine the municipal systems for the removal, treatment and disposal of the solid wastes produced in the jurisdictional areas and to draw the related projects;
- Carry on the project once approved by the Ministry;
- Plan, organize, remove, transport and treat the produced solid wastes.

1.5 Main research institutions

Universidade Nova de Lisboa—2 800 ALMADA
Centro Tecnológico de Saneamento Básico, Rua Antero de Quental, 44—1 100 LISBOA

2.0 Spreading of Information

2.1 Scientific and technical journals

APESB bulletin, issued four times annually and exceptionally 6 times.

2.2 Exhibition

In Portugal no specific exhibitions dealing exclusively with solid wastes are held. Companies dealing with solid wastes equipment or technology normally show it in general events and trade fairs.

2.3 Conferences

Conferences and seminars are normally organized by technical associations but not regularly.

2.4 University courses

Solid wastes are generally included in the course programme of Sanitary Engineering delivered by the Universidade Nova de Lisboa. On the other hand, the Civil Engineering "curricula" encompass some generalities about the subject.

3.0 Municipal Solid Wastes Disposal

The available data and disposal systems are shown in Table 1.

Table 1 Disposal systems in Portugal

Treatment and disposal facilities	Number of plants	Capacity (t/day)	Percentage of disposed wastes (%)
Incineration	—	—	—
Composting	2	800	14.4
Landfills	19	870	15.6
Dumps	—	3970	70.0

The source of the landfill data is from an investigation carried out by the Direcção Geral de Saneamento Básico in 1984. As to dumps capacity, no correct information is available and therefore the above-mentioned number was obtained from the difference from the amount of assumed total disposed wastes and that disposed by the other two techniques. With respect to the landfills, Table 2 shows how they can be classified according to the daily treated quantities.

Table 2 Landfill classification according to the quantities of waste received

Number of plants	Received quantity (t/day)
7	20
6	20–50
5	50–100
1	100

4.0 New Treatment and Disposal Facilities for Municipal Solid Wastes

At present, a new trend is taking place to recover old dumps into landfills. Some disposal sites have been provided with liquor-collecting piping and lining have been taken into consideration. Recently some incineration plants have been planned and studies have been carried out concerning the financial aspects and gaseous emissions.

5.0 Generalities

In a recent study carried out by DGSB, the maximum production of urban wastes per capita in Lisbon with 0.670 kg/inhabitant/day, and the minimum in Castelo Branco was 0.410 kg/inhabitant/day. The assumed total production of solid wastes in Portugal is 2 000 000 t/year. At present, 35.38% of the municipalities have a landfill.

14

Spain

Surface: 504 750 km^2
Population: 39 000 000 inhabitants
Number of municipalities: 8030 of which:

- 2 with more than 1 million inhabitants
- 51 with 100 000–1 million inhabitants
- 235 with 20 000–100 000 inhabitants
- 7742 with less than 20 000 inhabitants

1.0 Addresses

1.1 ISWA National member

International Solid Wastes and Public Cleansing Association (ISWA) – Spanish Section, Múgica y Butrón, nº 10 – 48007 Bilbao – Spain. Tel. 91 445 24 62

- National representative and Executive Committee member: Dr. Ing. Julián Uriarte Jaureguizar

1.2 Government offices

- Ministerio de Obras Públicas y Urbanismo
  Dirección General del Medio Ambiente
  Nuevos Ministerios – Paseo de la Castellana – Madrid
- Ministerio de Industria y Energía
  Dirección General de Innovación Industrial y Tecnología
  Paseo de la Castellana, 160 – Madrid
- Ministerio de Agricultura, Pesca y Alimentación
  Instituto Nacional par la Conservación de la Naturaleza (ICONA)
  Plaza Infanta Isabel, 1 – Madrid
- Ministerio de Sanidad y Consumo
  Dirección General de la Salud Pública
  Paseo del Prado 18 – Madrid

The Central Government is responsible for legislation in the field of the environment. The coordination of the ministries involved is realized through the CIMA (Comisión Interministerial del Medio Ambiente), whose president is the Minister of Public Works and Town Planning (Ministro de Obras Públicas y Urbanismo).

Spain is divided into seventeen autonomous communities, and each one is responsible for enforcing the legislation and for the actions in the field of the environment in the territory of its responsibility.

1.3 Associations

- ATEGRUS—Technique Association for the Management of Urban Solid Wastes.
  Múgica y Butrón, 10, 48007 Bilbao – España.

1.4 Main research institutions

- Empresa Nacional Adaro de Investigaciones Mineras, S.A. (ENADIMSA).
  Dr. Esquerdo n° 138 – Madrid.
  It makes plans for the use of wastes with application of the new technologies.
- Instituto para la Diversificación y Ahorro de la Energía (IDAE).
  Paseo de la Castellana n° 95 – 28046 Madrid.
  Promotion and financing of plans which involve economy of energy.
- Instituto Geológico y Minero de España (IGME).
  Ríos Rosas n° 23 – 28003 Madrid.
  Involved in plans and programmes that have to do with wastes and with the environment. Maps for the orientation of the disposal of wastes.
- Instituto Nacional de Investigaciones Agrarias (INIA).
  José Abacal n° 56
  Survey of the ground and recycling of wastes. Compost of sludge from wastewaters treatment.

2.0 Spreading of Information

- Periodic publications by the MOPU (Ministerio de Obras Públicas y Urbanismo).
- Bulletin of ATEGRUS, four per year.

2.1 Exhibitions

No specific exhibition exists exclusively dealing with solid wastes. The subject is normally included in more general events concerning the environment or with activities of the town council. The most important among them, taking place regularly, are:

- TEM (Técnicas y Equipamientos Municipales)—Municipal Techniques and Equipment. Recinto Ferial de Exposiciones y Congresos – Month of May – Madrid.
- MUNICIPALIA—Lérida.
- PROMA—International Monographic Fair of Environmental Protection – Bilbao

2.2 Congresses and conferences

Usually, the above-mentioned events imply the activity of symposia and conferences. Apart from those, there are others which are held regularly. The most important of them are:

- International conference of municipalities—Madrid.
- Technique day's work of PROMA—Bilbao.
- Symposium of Andalusia on urban solid wastes—Cádiz.
- National symposium on recovery of energy from wastes—Soria.
- South American Symposium on solid wastes—Madrid.

As training courses, the most important are:

- ITSEMAP—Technological Institute of Security MAPFRE—Centre of Environmental Hygiene—Avila.

3.0 The Elimination of Urban Solid Wastes

The available data can be found in Table 1.

Table 1 Disposal of municipal solid wastes in Spain

Treatment system	Number of installations	Capacity of management (t/year) (per 10^3)	Distribution (%)
Dumping	—	8620	81.3
Composting	40	1500	14.2
Incineration	10	480	4.5

In Spain, the main system for eliminating the solid urban wastes is dumping, and more than a half of the total dumping is not controlled. This situation affects mainly the small municipalities.

Many of the existing composting plants are obsolete, too little and of a deficient quality. Not more than ten can be really considered as composting plants with an adequate level. Generally, composting is associated with dumping as a system of elimination of wastes.

Incineration is absolutely stagnant, and apart from two small new plants of limited capacity, there has not been any further development.

4.0 New Plants and Systems of Elimination of Urban Solid Wastes

In the last five years, there has been very little achievements. The most important ones are the new composting plants with recovery systems more or less sophisticated. The results are uncertain because of technical difficulties and because the prices of the recovered products tend to fall. However, there has been an important implantation of the selective recovery of glass and making use of sludge coming from the treatment of residual urban waters with the first installations of management by composting. Furthermore, in various important landfills, biogas collection is beginning to be considered.

The actual tendency is directed towards the association of municipalities in the so-called "regional guiding plans", with big landfills well-equipped. An example of this is the plans developed on the island of Tenerife and in the central zone of Asturias.

15

Switzerland

Surface: 41 293 km^2
Population: 6 485 000 inhabitants
Municipalities: 3029 in total, of which:

- 5 with more than 100 000 inhabitants
- 4 with 50 000–100 000 inhabitants
- 87 with 10 000–50 000 inhabitants
- 336 with 3000–10 000 inhabitants
- 2597 with less than 3000 inhabitants

1.0 Legislation and Policy

The basic texts of Swiss law concerning wastes is represented by:

- Federal law relating to water pollution control, October 8, 1971. Principally the law, brought into effect in 1972, prohibits every direct or indirect discharge of dangerous liquids or substances into the waters. It calls for the treatment of all waste waters of households and industry.

  In the ordinance for waste water discharge (December 1975), quality criterias for surface water flows are set up as well as limits for a series of parameters that have to be met to guarantee the quality standards for effluents discharged into surface waters or into public sewers.

  Furthermore, special regulations for ground water protection and for the storage of dangerous liquids have been established.
- Federal law relating to the protection of the environment, October 7, 1983. According to this law, the protection of persons, animals and plants, their biological communities and habitats against harmful effects or nuisances, considers the principals of prevention, causality, cooperation and the principle of environmental impact assessment. Federal enforcement by concrete regulations governing the movement of special wastes (November 1986) have been enacted and an ordinance on wastes is under way. The legal basis for this ordinance in the law relating to the protection of the environment is rather weak and has to be completed. The federal council has already enacted ordinances on air pollution control, on dangerous substances and on noise abatement. These ordinances contain regulations that have to be respected by any waste disposal practice.

In 1986, the Federal Commission for Waste Management discussed and published principals of waste disposal for the near future (Leitbild für die Schweizerische Abfallwirtschaft). The new goal of waste management, waste disposal compatible with the environment, replaces a disposal practice focused on the lowest possible expense. The guidelines of the commission mentioned contend a series of technical standards as well as scientific and economic principals. Thus, disposal systems should produce only substances which can be recycled or materials that can be deposited in a final storage. The latter is only possible for substances and materials that do not pollute the environment (water, soil, air) to an intolerable extent, neither now, nor in the near or remote future.

Not suitable for final storage are fermentable and putrescible substances that can be found in municipal wastes. Organochemical wastes, such as waste oil, contaminated solvents and residues from chemical synthesis, have to be mineralized. Incineration is one of the most promising processes.

Slags, ashes and residues of the flue-gas treatment have to be transformed, together with other inorganic wastes, so that no environmentally relevant amounts of substances can be washed out. Without any doubt, special attention is needed for the heavy metals: a transformation into an insoluble "ore-like" material is required in case the wastes cannot be recovered. According to this concept, highly toxic or very reactive compounds are not deposited.

The Federal Office for Environmental Protection is preparing, together with scientists and authorities of the Cantons, the above-mentioned ordinance on wastes, containing stringent criteria for the choice of disposal methods as well as requirements for materials suitable for final storage.

2.0 Production and Composition of Wastes

Waste disposal, compatible with the environment, should be practised for liquid wastes as well as for solid wastes. A clear distinction between solid and liquid wastes or between municipal and special wastes is difficult to make as these wastes can partly be treated in the same facilities. Residues from these facilities have to meet the same standards for further treatment or depositing.

Collecting systems differ from region to region and even from some categories of wastes to others. Exact figures for the production of wastes are therefore very hard to get. The figures given underneath, most of them assessed have therefore to be interpreted with a certain reserve. However from these, one can get an idea of the problems waiting to be solved.

Switzerland has to deal with approximately 2.5 millions tonnes of municipal solid wastes. The composition of these wastes is given in Table 1.

Table 1 Composition of municipal solid wastes in Switzerland

	Percentage in weight (%)
Organic waste	29.4
Paper	20.8
Board	9.8
Plastics	13.4
Glass	8.7
Magnetic metals	4.7
Nonmagnetic metals	1.2
Paper-based compounds	2.4
Textiles	3.1
Minerals	1.5
Wood	4.3
Leather compounds	0.7

Beside these municipal solid wastes, collected from house to house, different materials are collected separately for recycling or special treatment, such as:

Glass

Refilling of glass bottles is very much encouraged and also practised, especially for beer, mineral water and wine.

Bottles that can not be refilled are collected in special containers according to their colour. About 140 000 t/year are recycled. Because of important quantities of imported green bottles, there is a surplus of green recycling glass, although the percentage of recycling glass in the production of new glass is very high (almost 100%).

Further information:
Vetro Recycling AG
6252 DAGMARSELLEN

Paper and board

In 1986, 465 000 t of paper were recycled, most of it for the production of board and only about 20% for the production of paper. Owing to the minor quality of paper collected from households (about 230 000 t/year) and the situation on the European waste paper market, a considerable surplus of collected paper led to the collapse of the prices. Some communities have started to give financial support to private organizations that have been collecting paper for years.

Further information:

- Verband Schweizerischer Industrielieferanten für Altpapier VSIA
  Neuengasse 39
  3011 BERN

- Verband der Schweizerischen Zellstoff-, Papier und Kartonindustrie ZPK
  Bergstrasse 110
  8030 ZÜRICH

- Förderverein Umweltschutzpapier FUSP
  Mühlensteg 8
  9470 BUCHS SG

Organic matter

During the last few years, communities and associations have started to collect organic matter separately and to run composting plants. Composting is seen as a good means to reducing environmental impact of landfills and to the recovery of valuable organic matter.

Textiles and metals

About 12 000 t of textiles for reuse or recovery are collected, mostly by private organizations. Most of the metal recycled is from used-cars. Together with other metallic items, about 260 000 t of iron are recycled. The amount of aluminium recycled is about 26 100 t/year, of which about 300 t are collected from private households.

Further information:

- Verband Schweizerischer Grosshandelsfirmen für Schrott Wengistrasse 42
  4502 SOLOTHURN

- Vereinigung der offiziellen Autosammelstellen-Halter der Schweiz (VASSO)
  c/o R. Sigg
  8108 BUCHS

Special wastes from households

Collected separately in special containers from small collection centres within the communities are: waste oil; residues of solvents, paints and pesticides; batteries, accumulators; vapour-lamps etc. (see Section on special wastes). In some communities, door-to-door collections are organized at intervals to collect such substances. Furthermore, appropriate disposal systems have to deal with the following quantities of wastes beside the municipal solid wastes:

- 3 million t of wastes from demolition, building sites, trade and small industries are deposited on landfills.
- 0.25 million t of sewage-sludge (dry matter). Some of the total 4.3 million t of sludge is used or dried, incinerated or deposited and mixed with chalk.
- 0.3 million t of special wastes have to be treated (see Section on special wastes).

In order to assess the possible environmental impact of the wastes, it is essential to know the fluxes of substances within the wastes. In this field, Prof. Dr. P. Baccini of the Swiss Federal Institute for Water Resources and Water Pollution Control (EAWAG) has published very valuable guiding contributions lately. (Einführung in die Methodik der Stoffflussanalyse, EAWAG, September 1987).

By measurements on several municipal solid waste incinerators, it was possible to get good indications about the substances contained in the wastes; the data are given in Table 2.

Table 2 Chemical characterization of municipal solid waste in Switzerland

	(ppm)
Organics (TOC)	350 000
Water	300 000
Iron	45 000
Aluminium	6500
Chlorine	7000
Lead	500
Cadmium	10
Chromium	60
Cobalt	6
Copper	400
Molybdenum	10
Nickel	40
Mercury	4
Zinc	1500

3.0 Collection Systems

Municipal solid wastes are collected either by communities or associations that mostly run an incinerator or are also responsible for a sanitary landfill.

The collection systems for recyclable or hazardous materials differ not only from region to region according to the disposal system, but also from one category of wastes to another.

Solid wastes from households and similar wastes from shops, administration buildings, hospitals, small industries etc. are collected by the door-to-door system. Also in some communities, organic matter, paper and glass are collected by this system. In other communities, people can deposit their wastes at collection centres, where there are also containers for other materials such as aluminium, waste oil, solvents, batteries, vapour lamps. Textiles are collected by private organizations, sometimes paper and board too.

Further information:

- FES/ORED
  Schweizerischer Städteverband
  Junkerngasse 56
  3000 BERN 8

- Aktion Saubere Schweiz
  Kernstrasse 31
  Postfach 5
  8026 ZÜRICH

- Schweizerische Vereinigung für Gewässerschutz und Lufthygiene
  Schaffhauserstrasse 125
  8057 ZÜRICH

4.0 Waste Treatment and Disposal

In Switzerland, the realization of urgently-needed facilities for the treatment and final disposal of all kinds of wastes is made very difficult, delayed or even blocked up by the current mistrust of people. This is due to the formerly practised and still prevailing mismanagement of facilities which do, often enough, not correspond to the latest state of technology.

This is especially true for the disposal facilities of the so-called special wastes, i.e. for wastes that require special treatment because of their chemical and physical properties and cannot be disposed of together with municipal solid wastes. In our country, the existing treatment capacities for such wastes are insufficient.

4.1 Incineration

From the 2.5 million t/year of municipal solid wastes, more than 80% is incinerated in 39 facilities. From about three-quarters of the incinerated wastes, the heat is recovered. Up to the end of 1987, only six facilities (corresponding to approx. 20% of the incinerated wastes) were equipped with scrubbers for the treatment of flue-gas. Not yet solved in some of the facilities, is the disposal of the slags, that has partly been used for road construction, partly deposited in landfills which are insufficiently equipped. The disposal of filter ashes and residues of the flue-gas treatment should be solidified and deposited correctly (see report "Behandlung und Verfestigung von Rückständen aus Kehrichtverbrennungsanlagen", Schriftenreihe Umweltschutz Nr. 62, Federal Office for Environmental Protection, 3000 Bern).

4.2. Sanitary landfill

Together with about 500 000 t of municipal solid wastes, about 3 million t of wastes from demolition, building sites and small industries are deposited. Unfortunately, considerable quantities of such materials are not deposited in sanitary landfills corresponding to the latest state of technology, but end up in small landfills that are run by small communities, often are in unsuitable sites and not controlled. The new ordinance on wastes will set standards to improve the situation.

4.3 Composting plants

Actually in more than 30 plants, compost that meets the quality standards set up in the ordinance on dangerous substances is produced. Compost made out of the organic material collected together with other municipal solid waste is no longer produced in quantities, as its quality is insufficient, i.e. its heavy metal content is far too high.

5.0 Special Wastes

5.1 Legal basis

Movement and control of special waste are governed by a federal ordinance "Verordnung über den Verkehr mit Sonderabfällen", dating from April 1, 1987. Its main features are:

- Categories of special waste are defined by listing in an appendix to the ordinance;
- Special wastes may only be transferred to persons or companies

running appropriate treatment facilities and being in possession of a state authorization;

- The Cantons (state of the Confederation) are responsible for the control of the treatment facilities and of any movement of special waste;
- Any exportation of special wastes must be permitted by the Federal Authorities and has to be notified to the authorities of the transit and receiving countries.

5.2 Quantities of special wastes

At present, only estimations of the annual quantities of special wastes are available. The first statistics, based on the data obtained by the control according to the ordinance mentioned above, will not be published before the end of 1988. About 300 000 t of special waste are produced annually (Table 3).

Table 3 Estimated annual quantities of special wastes (t/year)

Aqueous solutions; inorganic	34 000
Inorganic sludges (dry weight)	11 000
Inorganic dusts and solids	40 000
Batteries, accumulators, vapour lamps	19 000
Solvents	53 000
Fats, oils, lubricants	60 000
Organic sludges, paints	75 000
Emulsions	23 000
PCB-containing transformers and capacitors	500
Chemicals in small quantities (laboratories, households)	200

5.3 Installations for special waste treatment

Aqueous solutions with inorganic composition are treated by neutralization, precipitation and filtration, in existing facilities at industrial plants or by specialized waste-treating companies.

After solidification, e.g. by cement, the residues, aswell as other inorganic sludges or dusts, may principally be deposited in controlled landfills. Appropriate sites are under investigation. Currently, most of these special wastes, as well as discarded batteries and vapour lamps, are exported.

Controlled landfill of organochemicals is not considered as an appropriate means of disposal any longer. Organochemical wastes should be burnt in high temperature incinerators. Today's capacity in Switzerland for incineration of special waste is about 35 000 t (four installations). Other, unproblematic materials, like oils or nonhalogenated solvents, might be used as fuel in cement kilns.

Projects are under way for additional high temperature incinerators; partly by industrial companies, partly by cantonal authorities.

6.0 Addresses

Federal Office for Environmental Protection
BUS
Hallwylstrasse 4
3003 BERN

Institutes

- Swiss Federal Institute for Water Resources and Water Pollution Control (EAWAG)
  Prof. Dr. P. Baccini
  Head of the Dep. of Waste Management and Material Balances
  8600 DÜBENDORF
- Institut du génie de l'environnement
  Directeur Prof. L. Y. Maystre
  EPFL Ecublens
  1015 LAUSANNE
- Institut de Botanique de l'Université de Neuchâtel
  Chantemerle 22
  2000 NEUCHÂTEL

16

Sweden

Surface: 449 945 km^2
Population: 8 399 000 by January 1, 1988
Counties: 24
Municipalities: 284 of which:

- 1 (Stockholm) with more than 600 000 inhabitants
- 1 (Gothenburg) with 400 000–600 000 inhabitants
- 1 (Malmö) with 200 000–400 000 inhabitants
- 8 with 100 000–160 000 inhabitants
- 26 with 50 000–100 000 inhabitants
- 183 with 10 000–50 000 inhabitants
- 64 with less than 10 000 inhabitants

1.0 Legislation

The legislative framework for solid waste management is given by the following acts:

- The Environment Protection Act (1969) and the Environment Protection Ordinance (1981)
- The Cleansing Act and the Cleansing Ordinance (1979)
- The Ordinance on Hazardous Waste (1975)
- The Vehicle Scrapping Act and the Vehicle Scrapping Ordinance (1975)

As a result of the first "Public Cleansing Act" which came into force in 1972, responsibility for public cleansing was placed on the municipalities, as regards collection and transport of domestic waste. In the revised "Cleansing Act" from 1980, it is moreover prescribed that the municipalities were obliged to ensure that treatment plants for domestic waste are available.

In the "Cleansing Ordinance" from 1985, it was stipulated a responsibility on the statute book for all municipalities to collect and transport hazardous waste from the industry. In the "Ordinance on Hazardous Waste", 12 different kind of wastes are defined as hazardous wastes. The ordinance also regulates the conditions which must be fulfilled to get permission to transport, treat and export hazardous waste.

With the exception of the "Ordinance on Hazardous Waste", there are no stipulations on the statute book as to the management of industrial waste.

However, the municipalities may, with the support of the stipulations in the "Cleansing Act" increase the municipal collection responsibility, so as to cover wastes other than domestic waste. So far about 40 municipalities have made use of this possibility.

2.0 Policy

The Swedish Government has during 1986–87 given the following guidelines for future solid waste management in Sweden:

- The quantity of waste should be reduced by measures taken on both the production side and the consumption side. This should be a basic and common community task. Also, the producers have a responsibility for the protection of the environment.
- Produced waste should be reused or recovered as far as it is practically and economically possible.
- Waste that must be processed should be handled and disposed of in such a way that negative effects on the environment will be minimized in both the short- and long-term. Stringent environmental protection requirements must thus be established.
- The handling of hazardous waste must be improved. Substances and products which, after being mixed with other types of waste, cause special damage to the environment should be handled separately.
- There must not be any one-sided emphasis on a particular method of waste treatment, or a situation which forces one particular method to be adopted. Alternative methods should be developed, tested and tried.

In 1985 the Swedish Association of Public Cleansing and Solid Waste Management presented its views on waste management in the future in a report titled "Future Waste Treatment in Sweden—A Combination of Methods to Conserve Resources and Improve the Environment". In summary the following was stated:

> The management of the Swedish waste mountain must be based, both now and in the future, on a combination of different methods, where the choice of method must put environmental protection first. There must not be any contradiction between the different methods; instead they should be regarded as complementing each other. Only in this way the waste problem can be solved in an optimal fashion. Large-scale and small-scale methods must go hand in hand and complement each other.

In the future, it would appear that emphasis within solid waste management should be placed on the following:

- Recovery of material in households and industries, involving direct action by residents and industries in cooperation with those responsible for waste collection;

- Improved and rationalized collection and transporting of different kind of wastes;
- Thermal treatment of waste with energy recovery (provided that the environmental consequences of emissions can be kept under control even in the future);
- Landfilling, which always will be needed regardless of other methods utilized, for non-reclaimable materials, residual products from other treatment etc.;
- Continued research and development on central mechanical separation and composting of waste;
- Development of large-scale methods for production of biogas out of organic waste;
- Increased utilization of material and energy resources in industrial waste. The conservation of resources should chiefly take place in the industries themselves, by using low-waste technology and by internal recovery of the waste products;
- Improved collection and treatment of hazardous wastes from the industry.

The representatives of waste management in Sweden are trying to minimize the environmental effect of produced waste by means of far-reaching environment-protective measures and *not* to cause further damage. The causes of environmental disturbances are usually to be found far back in the chain—in production and consumption.

3.0 Administration

The environment protection field in Sweden is broad. The environment protection authorities have an extended contact surface with other authorities. Even though the National Environmental Protection Board is the central government authority in the field of environment protection in Sweden, it should be observed that the majority of the decisions that govern the work of environment protection are made by bodies other than the Environment Protection Board, mainly the Government, the Franchise Board for Environment Protection, the Products Control Board, the county administrations and the municipalities.

At Government level, responsibility for environmental questions is largely vested in the Ministry of the Environment and Energy.

The National Environmental Protection Board is the central administrative authority in the environmental sector, i.e. the authority that is supposed to execute the decisions of the Parliament (Riksdagen) and the Government in this sphere, keep track of new developments, and propose necessary measures to the Government.

The main responisibility for environment protection at the regional level rests with the county administrations. They are the regional supervisory authority under the Environment Protection Act and theHealth Protection Act.

Local responsibility for nature conservancy rests with the municipalities. The municipal environment and health protection committees, as well as the building committees, have important environment protection functions to fulfill.

The executive solid waste management is carried out by local authorities and by private contractors.

4.0 Categories, Amounts and Composition of Waste

4.1 Categories

The following categories of waste are collected or brought to final treatment or disposal in the municipalities:

- Household waste
- Industrial waste, including construction and demolition waste
- Hazardous wastes
- Sewage sludge
- Ashes and slags from power plants

Trade waste is included in both household waste and industrial waste.

Not included in industrial waste are wastes from agriculture, forestry, mining, fishery etc. (all together appr. 65 Mtons/y).

Not included in household waste are reused and recovered materials such as bottles, cans, newspapers and cardboard.

4.2 Amounts

There is a yearly production of household waste in Sweden of approx. 2.5 million tonnes, that means approx. 300 kg/person/year or approx. 750 kg/household/year. Depending upon social conditions, living structure, geographical location, the amount/person/year varies between 150–450 kg. The total amount of household waste is not supposed to increase. On the contrary, it is supposed to decrease due to an increased recovery of different materials out of waste.

The yearly production of industrial waste brought to final disposal and treatment in the municipalities is estimated to be about 4.5 million tonnes. Due to the varying industrial structure from municipality to municipality the amount of industrial waste also varies to a great extent.

The total amount of hazardous waste is most uncertain, but is estimated to be about 0.5 million tonnes. The figure indicates all hazardous waste produced in Sweden, including the internal recovery within and between different industries.

Yearly received amount of sewage sludge: 0.4 million tonnes as fresh weight (25% dry solids).

Yearly received amount of ashes and slags: 0.6 million tonnes.

4.3 Composition

Average composition of household waste is given in Table 1.

Table 1 Average composition of Swedish household waste

Fraction	Percentage of weight (%)
Paper	35–45
Plastics	8–10
Garbage, garden waste	25–35
Textile, leather, rubber	2–4
Metals	2–4
Glass	6–8
Wood	1
Miscellaneous	6–8

The high content of paper, an increasing content of plastics and a decreasing content of metals result in a high calorific value, 10–12 MJ/kg.

There are only figures from a few places in Sweden upon the composition of industrial waste. The composition varies widely from place to place depending on size and type of industry. In general about 50% of the industrial waste is estimated to be burnable with a calorific value of approx. 15 MJ/kg.

5.0 Collection of Waste

5.1 Organization

The collection of household waste is a municipal responsibility. About 50% of the amount of household waste is collected by the municipalities and 50%

by private contractors. The contractors are doing the collection on commission from the municipalities. About two-thirds of the municipalities are using contractors. About one-third is doing the collection on their own. A few large contractors are operating regionally or even nationwide, but most contractors are small and operate only locally. Most of the large cities operate their own waste collection department.

Most of the household waste from the apartment houses is collected from the yard or from special refuse storage rooms in the basement or on the ground floor. From the villas, the waste is mostly collected from a special small refuse room in the front wall of the house or from a special refuse box at the gateway. After collection, the empty receptacles are returned to the same location. Vehicles are generally manned by two men, both taking part in the loading.

In low-density areas in the countryside and in areas with summer houses, the households themselves can often deliver the waste directly to centrally-located containers.

Weekly collection of household waste is almost universal. Only in the high-density areas of the largest cities is more frequent collection sometimes necessary. In the low-density areas in the countryside, collection is sometimes necessary only once a fortnight.

Most collection schemes include a regular or by-request collection of discarded furniture and other bulky waste from the households.

In the cities, an important proportion of the waste from offices, shops etc. is collected under the household collection scheme. However, the collection of commercial and industrial waste is generally left to private enterprises and carried out by direct contracting with the individual producing companies.

Industrial waste and hazardous waste is mostly collected by private contractors, but in some cities also, the public cleansing departments are collecting industrial wastes.

5.2 Refuse receptacles

Since its introduction in the 1960s, the two-ply paper bag (125 l to 350 l) has held a remarkably high share of the market for household refuse receptacles. About 75–80% of the household waste is collected in paper or plastic bags. Approx. 85% of the bags are made of paper and 15% of plastic. At present, about 20–25% of the household waste is collected in plastic or steel bins, in containers or, to a small extent, through vacuum systems/pipelines.

In recent years, the modern two-wheeled plastic bins (190 l, 240 l and 400 l) are gaining their share of the market and in some places, replace the use of paper bags. Cheaper plastic bags have also been gaining their share of the market in recent years.

Bags, paper or plastic, are always placed in special stands. For installation under refuse chutes, special equipment including automatic shifting devices is sometimes used in high-rise buildings.

Large containers of the roll-on/roll-off type or skips of winch/rear loading type are mainly used for industrial/commercial waste and building waste but to a modest scale also for household waste from high-rise buildings and for garden waste or bulky household waste. Also front loaders are used for collection of industrial waste, especially in the large cities.

5.3 Vehicles

The distance between the collection area and the treatment plant is usually short in Sweden. In about 60% of all high density areas, the distance is shorter than 10 km and only in 10% is the distance longer than 20 km.

There are at present about 1800 vehicles for collection of household and industrial waste in Sweden. Most of the vehicles are of the compacting type, with a (statistically) medium volume of 15 m^3 and a loading capacity of about 6 t. The yearly demand of new compacting vehicles in Sweden is about 175–200.

For collection of household wastes, there is a growing interest for smaller and lighter vehicles, often in combination with transfer stations. These lighter vehicles are of special interest at collection in narrow streets, in the countryside and at collection of recovered materials. For collection of industrial waste, there is a growing interest and demand for large compacting vehicles with capacities to handle containers from 2–10 m^3. A number of vehicles for combined refuse collection and recycling are in use and under development.

5.4 Automized collection systems

In new and large residential areas and in larger buildings (hospitals etc.), automized collection systems such as vacuum systems with suction through pipelines are to be seen. Also mobile suction systems are under development and test.

6.0 Recovery, Treatment and Disposal of Waste

About 2.5 million tonnes of household waste are delivered every year for final treatment or disposal in Sweden, i.e. the amount after recovery of different materials in the households.

6.1 Recovery

Concurrently with the separation of waste at large facilities, attempts have been made to divide household waste into fractions and to segregate products that are hazardous to health and the environment prior to collection (source recovery). Increasingly, these attempts have been made during recent years. The main objectives of these efforts are to increase the amount of material that can be recycled, to improve the way in which we handle environmentally hazardous waste components, and to produce cleaner waste that is better suited for subsequent treatment and disposal.

Three different systems are today practised in connection with source recovery in the households:

- Source recovery with collection integrated with the ordinary collection of refuse. The households separate the waste in 1–4 fractions—newspapers, glass, small hazardous batteries and ordinary refuse or alternatively in a wet and a dry fraction;
- Source recovery with collection segregated from the ordinary collection of refuse, i.e. collection of the separated materials in accordance with a certain schedule;
- Source recovery with collection of the recovered materials at centrally located containers, predominantly for paper and glass.

The best example in Sweden of a successful source recovery is the recovery of newspapers and journals. The activity is nationwide. The average recovery amounts to 280 000 t (33 kg/year/person), corresponding to a rate of recovery of 50%.

At present, about 22 000 t of glass from the households are recovered per year, corresponding to about 2.6 kg/person/year and a rate of recovery of 15%. "Glass igloos" are provided in many municipalities. More than 200 municipalities (of 284) are taking part in the recovery of glass.

To reduce the content of heavy metals in waste, the government has drawn up a bill which sets forth a programme of measures to control the discarding of environmentally hazardous batteries. In the first phase of this programme, the amount of mercury released into the environment from batteries will be reduced by at least 75%. To achieve this objective, the government has taken a number of measures, including information, collection and marking. Information, final treatment and marking are being financed by means of fees. The collection and the dividing of the collected batteries into two fractions (hazardous and nonhazardous) are financed by the municipalities. According to a governmental ordinance on hazardous batteries, such batteries must not be mixed with household waste.

Since 1987, all Swedish municipalities are taking part in a campaign to collect small hazardous batteries. The information includes advertisements in newspapers, on TV, on buses etc. It also includes a special information to all schools. The households can deliver the batteries in special battery-boxes (located on streets, markets etc.), to shops where batteries are sold and sometimes in connection with the collection of ordinary household waste. At present, about 50% of the mercury content in sold batteries is returned.

In 1982, the Parliament made a decision on the recovery of aluminium cans used for beverages, whereby at least 75% was to be recovered not later than 1985. A special company was formed to run a deposit system. The deposit system was introduced in 1984. During 1985, 66% of the cans were recovered. The recovered amount during 1986 was approx. the same. To achieve the recovery level of 75%, the consumer's economic incentive was doubled during 1987, to SEK 0.50 per can. For the moment, about 85% of the cans are returned. Also for glass bottles for beverages, a special deposit system is used. Each bottle is on average returned and reused 33 times.

Waste recovery has also reached significant levels with regard to certain types of industrial materials. The systems that have been developed within both municipalities and industries to recover residuals and to recycle materials, have eliminated any increase in the amount of waste generated during the last ten years.

6.2 Treatment and disposal

Treatment and disposal of solid waste is summarized in Table 2, by relative measures for certain waste categories.

Table 2 Waste treatment and disposal in Sweden in 1987 (after source recovery; percentage of weight, %)

Category	Land application	Incineration	Separation/ compost	Landfilling
Household waste		55%	10%	35%
Industrial waste		5%		95%
Sewage sludge	60%			40%

Incineration

Thermal treatment of waste combined with energy recovery is widely used in Sweden. At present, more than 50% of the municipal solid waste produced in the country is incinerated in about 25 waste-to-energy plants. In recent years, there has been a growing interest in industrial waste utilization. Today, about 10–15% of the combustible fraction of solid wastes from industries is used for energy production.

Table 3 shows the development during the 1980s and the potential concerning incinerated amounts and energy production.

Table 3 Incinerated amounts of waste and energy recovery

Year	Incinerated amounts of waste (million tonne) Household	Industrial	Energy recovery (TWh) Total
1980	0.72	0.14	1.4
1983	0.91	0.17	2.3
1986	1.29	0.24	3.4
Potential	2.0–2.5	1.5–2.0	9–11

During the last six-year-period, the incinerated amount of waste has nearly doubled. As indicated in the table above, the average specific energy recovery has increased by about 30% during the same period. This is mainly a result of a gradually decreased amount of heat being cooled off without recovery. At present, 97–98% of the energy produced is utilized. A small proportion of the recovered energy is used for steam and electricity production, roughly 1 and 2% respectively. The main part is utilized for production of hot water for district heating purposes. About 10% of the district heating requirement in the country is covered with heat from waste incineration facilities. For some cities, this proportion reaches 30–40%.

The environmental impact from waste incinerators has shown great concern in recent years. The Swedish regulation for control of flue-gas emissions is one of the most restricted of its kind in the world. The most important emission requirements are shown in Table 4.

Table 4 Emission requirements from incinerators

Parameter (mg/nm^3 dry gas, 10% CO_2)	Emission requirement	
Dust	20	(monthly mean value)
HCl	100	(monthly mean value)
Hg	0.08	(when inspected)
TCDD*	0.5–2.0	(existing plants)
	0.1	(new plants)

* Dioxins measured as TCDD-equivalents according to Eadon.

The Government decided in 1987 that the new emission standards must be met before 1992. This will probably result in waste incineration facilities being equipped with advanced flue-gas treatment. The waste incineration plants are traditionally equipped with dust removal processes, i.e. electrostatic precipitator, fabric filter etc.

The installation of advanced flue-gas cleaning systems is increasing rapidly. The first dry method application (lime addition) was already installed in 1981. At present, there are six or seven facilities in operation. The first flue-gas condensation scheme began to operate in 1986. The present situation and the calculated situation in 1989, based on made orders, concerning advanced flue-gas treatment are shown in Table 5.

Table 5 Advanced flue gas treatment

Advanced flue gas treatment method	Percentage of incinerated amounts of waste	
	Year 1987	Year 1989
Dry	30	45
Wet	15	35
Total	45	80

Compehensive research and development work has been carried out in recent years to improve waste incineration and reduce the emissions of pollutants. The DRAV-project—an operational study of waste treatment methods—was conducted during the years 1981 to 1985 by the Swedish Association of Public Cleansing and Solid Waste Management (RVF) in cooperation with the National Swedish Environmental Protection Board (NSEPB). Another investigation—Energy from Waste—was carried out by NSEPB and the National Swedish Energy Administration and published in 1986. Both studies concluded that waste incineration with energy recovery is technically and economically a sound waste handling method, as long as the environmental aspects are taken into consideration.

Basic research efforts at present are primarily focused on:

- Waste processing to obtain higher-quality waste fuel;
- Further optimization of the combustion process;
- Formation mechanisms concerning mercury compounds and chlorine aromates;
- Recycling and disposal of slag, fly ash and other residues.

The Swedish Association of Public Cleansing and Solid Waste Management (RVF), in capacity as the solid waste management organization in

Sweden, is cooperating with universities, governmental agencies and private companies in different research and development projects. The "Waste Incineration Group", an executive committee within RVF, is conducting projects within smaller projects groups. Some of the ongoing projects include:

- Exposure to heavy metals and organic compounds to plant operating staff;
- Experiences from flue-gas condensation facilities;
- Experiences from incineration of industrial waste;

Table 6 Incineration plants in Sweden

Plant location	Type	Capacity (t/year)	Established (year)	Flue-gas cleaning
Avesta	Grate	37 900	1980	ESP
Bollmora	Grate	3500	1967	ESP
Bollnäs	Fluid-bed	11 500	1983	FF
Borlänge	Grate	10 000	1983	ESP
Borås	Grate	26 500	1986	ESP
Eksjö	Fluid-bed	4500	1979	WET
Göteborg	Grate	295 000	1972	ESP
Halmstad	Grate	45 200	1972	WET
Huddinge	Grate	2800	1972	ESP
Karlskoga	Grate	9000	1986	ESP
Karlstad	Grate	14 800	1986	LIME/FF
Kinda	Grate	17 500	1984	ESP
Kiruna	Grate	8500	1985	ESP
Köping	Grate	38 000	1972	ESP
Lamdskrona	Fluid-bed	7000	1983	FF
Lidköping	Fluid-bed	11 200	1986	LIME/FF
Linköping	Grate	200 000	1981	LIME/FF
Malmö	Grate	215 000	1973	LIME/FF
Mora	Grate	18 000	1981	ESP
Stockholm	Grate	200 000	1970	LIME/FF
Sundsvall	Fluid-bed	7900	1984	ESP
Trollhättan	Grate	15 000	1968	FF
Umeå	Grate	75 000	1970	ESP
Uppsala	Grate	250 000	1961	FGC
Västervik	Fluid	12 000	1984	FF

ESP: Electrostatic Precipitator, FF: Fabric Filter, LIME: Lime reactor, WET: Washing, FGC: Flue Gas Condensation.

- Improvements of the fluid-bed technology;
- Combined heat and power production from waste;
- Emission levels and removal technologies concerning SO_x and NO_x contents in flue-gas;
- Training courses for plant operating staff;
- Utilization of waste incineration slag as road construction materials.

Central separating and composting

Almost 10%, or 200 000–250 000 t, of Swedish household waste is treated in central plants for separation and composting. At present, 12 plants are in operation.

Against the background of governmental grants and an assessed market for materials recovered from waste, 15–20 new plants for separation and composting of waste were built from 1977–1982. Since the technology was new and untested for Swedish conditions, all plants quite naturally, ran into operational problems in their initial phases. Most of these problems were solved after one or two years of operation.

For the moment there is a yearly production of RDF of approx. 90 000 t, unpelletized or pelletized. About 75 000 t are sold or used in some way through agreements with different buyers.

There is a yearly production of compost of approx. 50 000 t. Only 10 000 t are sold. The rest is mostly used as cover material on the landfills. The difficulties to sell the compost are mainly due to the heavy metal content.

No new plant for central separation and composting is planned for the moment.

Table 7 Composting plants in Sweden

Plant location	Yearly amount (t)	Products	Established
Borlänge	30 000	RDF, Compost	1980
Falkenberg	25 000	RDF, Compost	1981
Karlshamn	15 000	RDF, Compost	1981
Landskrona	18 000	RDF	1987
Ludvika	10 000	RDF, Compost	1979
Stenungsund	21 000	RDF, Compost	1983
Strömstad	8000	RDF, Compost	1980
Sundsvall	25 000	RDF, Compost	1982
Uppsala	10 000	RDF	1978
Kovik	30 000	RDF, Compost	1982
Växjö	26 000	RDF, Compost	1974
Österåker	20 000	RDF, Compost	1978

Landfilling

Landfilling is widely used in Sweden. There are about 400 sanitary landfills in operation. 4000 completed sites throughout the country have been identified. There is a strong tendency towards larger landfills, creating increased possibilities for enhanced environmental protection and separation of materials at the landfills (i.e. wood fuels, scrap materials, oil wastes, hazardous wastes, etc.).

At present, about 6 million tonnes/year of different categories of waste are landfilled (Table 8).

Table 8 Landfilled amount of waste

Category	Amount (million tonnes/year)
Household waste	1.0
Industrial waste	4.0
Sewage sludge	0.4
Slags, Ashes	0.6

At present, 11 landfill gas utilization facilities in operation produce 50 GWh/year (3–5% of the usable energy potential) mainly for heating purposes (district-heating, green houses, RDF-drying). Many municipalities are planning for gas extraction (energy as well as environmental aspects).

Leachate (the most important environmental problem) is cotreated with municipal wastewater at one-quarter of the sites. On-site leachate treatment (pond aeration, infiltration, soil-plant application) is used at some 30 landfills.

Sewage sludge

The annual production of sewage sludge in Sweden is estimated to 0.25 millions tonnes dry matter (approx. 30 kg/person/year). All citizens are connected to secondary waste water treatment and about 80% also to a tertiary treatment step (chemical precipitation with phosphorus removal).

All sludge is stabilized (mainly by anaerobic digestion) and mechanically dewatered (centrifugation, filter-pressing). Digested gas is burned and used for internal heating purposes (digester, buildings).

Digested and dewatered sludge is landfilled (40%) or used in agriculture (60%). A sludge application rate of five t dry matter per hectare during a five-year period (1 t DM/ha and year) is recommended.

Guidelines for metal content in sludge utilized in agriculture are presented in Table 9.

Table 9 Guidelines for metal content in sludge utilized in agriculture

Metal	Recommended value (ppm dry solids)
Cadmium (Cd)	4*
Mercury (Hg)	5
Nickel (Ni)	100
Chromium (Cr)	150
Lead (Pb)	200
Copper (Cu)	600**
Zinc (Zn)	1500

* Limit value.
** Up to 1200 for soils needing addition of Cu.

The content of trace organics in sludge applied to farming land is frequently discussed. At present, land application of sludge is the most viable method for sludge disposal.

7.0 Hazardous Wastes

It was not until 1975, with the "Ordinance on hazardous wastes" that a comprehensive concept of such wastes—hazardous wastes—was introduced. This ordinance regulates the treatment of hazardous wastes, and covers the following points:

- Definition (categorization) of hazardous wastes;
- Obligation for anyone whose commercial activities involve the generation of such wastes to provide statements concerning the type, composition, amount and treatment of hazardous wastes;
- Transportation of hazardous wastes;
- Final disposal/treatment of the wastes;
- Export of hazardous wastes;
- Supervision and observance of the ordinance.

The ordinance does not apply to hazardous wastes which are included in household wastes.

A Government treatment company, SAKAB, was established in its present form in 1974 and is responsible for the necessary resources for the treatment of hazardous wastes on a national basis, in accordance with the ordinance on hazardous wastes.

All transportation of hazardous wastes is subject to current regulations for the form of transportation in question. For example, in case of land transportation the vehicle must carry a plate which in accordance with ADR states the primary risk and classification of the load, as well as documents which state the chemical composition etc., of the goods—a complete declaration of the load, in other words. If hazardous wastes are to be transported out of the country, a special permit is required.

The great majority of hazardous wastes in Sweden are transported by road, though some is by rail. Liquid waste is transported in tanks and sludge suction trucks or by rail tank transports. Other waste is normally transported in barrels, standard containers or environmental boxes, for example. All transports must be authorized and the necessary transport documents must accompany the transport.

The generator shall render an account of the type, quantity and composition of hazardous wastes generated as a result of commercial activities. On transfer of delivery, a declaration of the waste is to be provided on a special form. The transporter should be authorized, and all the necessary documents to be carried and plates displayed during transport. Treatment facilities should have the necessary permits required by current legislation.

A guideline list of hazardous wastes is published by the National Swedish Environmental Protection Board. This list constitutes the basis for application of the "Ordinance on hazardous wastes". At present, the list contains 12 main groups of waste which are defined as hazardous wastes. The main categories are indicated in Table 10.

Local authorities will be required by law to collect hazardous wastes. In such cases as the local authority itself does not carry out the work, a

Table 10 Quantities of hazardous waste in Sweden 1980 (quantity in t)

• Oil wastes	183 000
• Solvent waste	34 600
• Paint and varnish waste	20 700
• Adhesive waste	10 800
• Acid or alkali waste	72 300
• Cadmium-contaminated waste	480
• Mercury-contaminated waste	440
• Waste containing compounds of antimony, arsenic, barium, beryllium, led, cobolt, copper, chrome, nickel, selenium, silver, thallium, tin or zinc	120 000
• Cyanide-contaminated waste	3500
• PCB-contaminated waste	31
• Pesticide waste	580
• Laboratory waste	41 000

contractor may be engaged. Such contractors must be licensed by the county administration. Transportation may be carried out by both public (local council) or private transporters. Private transporters are required to be licensed.

A comprehensive picture of the management of hazardous wastes in Sweden is provided in Table 11, both in the form of distribution by percentage of different waste treatment methods and according to waste managers.

Table 11 Management of hazardous wastes in Sweden in 1980

	(%)
Method	
Incineration	20
Recycling/conversion	24
Other use	2
Landfill	40
Sewerage	12
Other methods	1
Storage (management not yet solved)	1
Manager	
Own management by generator (industries etc.)	36
SAKAB	10
Other disposal companies	33
Municipalities	21

In certain municipalities, a small truck drives around the various areas, according to a schedule, and collects household hazardous wastes. The general public therefore have the opportunity of disposing of their hazardous wastes where they live. Old unused medicines and broken thermometers may be handed in at chemists, who then forwards them.

The general public may also, in certain municipalities, book collection of hazardous wastes at the next collection of ordinary household waste. This must be well packaged and clearly marked. This procedure is followed in the countryside.

8.0 Addresses

8.1 ISWA National member

The Swedish Association of Public Cleansing and Solid Waste Management (Svenska Renhållningsverks-Föreningen, RVF)

Address Östergatan 30
211 22 MALMÖ
Sweden
Tel. 0 40 104045
Telefax: 0 40 971094

RVF is represented in the ISWA-Governing Board, in the working group "Hazardous Wastes" and in the working group "Sanitary Landfilling".

8.2 Government offices

The Swedish Ministry of the Environment and Energy (Miljö- och Energidepartementet)

Address STOCKHOLM
Sweden
Tel. 0 7631000

The National Swedish Environmental Protection Board (Statens Naturvårdsverk/SNV)

Address P.O. Box 1032
171 25 STOCKHOLM
Sweden

The National Swedish Energy Administration (Statens Energiverk/STEV)

Address 117 87 STOCKHOLM
Sweden

8.3 Technical and professional associations

The Swedish Association of Public Cleansing and Solid Waste Management (Svenska Renhållningsverks-Föreningen/RVF)

Address, telephone and telefax see "ISWA National Member" above.

RVF is an organization for the Swedish municipalities in questions concerning public cleansing and solid waste management—collection, transportation and treatment of different kinds of waste. The office in Malmö is dealing with questions concerning technique, economy and environmental impact in connection with public cleansing and solid waste management. Within the association there 7 different working groups:

- "Collection, Transportation of Solid waste";
- "Incineration of Solid Waste";
- "Central Separation and Composting of Solid Waste";
- "Sanitary Landfilling";

- "Hazardous Waste";
- "Industrial Waste";
- "Public Cleansing of Streets, Markets, Pavements etc.".

RVF was founded in 1947.

8.4 Main research institutions

REFORSK Foundation (Stiftelsen REFORSK)
Address Östergatan 30
211 22 MALMÖ
Sweden
Tel. 0 40 129880

REFORSK is a foundation for cooperation between national and local governments and industries in research and development within the field of waste management and recovery technology. The following major points may be singled out from the general programme:

- technical development
- material recovery
- energy recovery
- systems studies
- build-up of relevant knowledge.

REFORSK is owned by the following bodies:

- The Federation of Swedish Industries (Sveriges Industriförbund)
- The Swedish Association of Municipal Authorities (Svenska Kommunförbundet)
- The Swedish Association of Public Cleansing and Solid Waste Management (Svenska Renhållningsverks-Föreningen)

REFORSK was founded in 1984.

8.5 University institutions

Division of Residual Products Technology, University of Luleå (Institutionen för Restproduktteknik, Luleå Tekniska Högskola)
Address: 951 87 LULEÅ
Sweden
Tel. 0 920 91000

Division of Wastes Biology, Department of Soil Sciences Uppsala (Avdelningen för Avfallsbiologi, Lantbruksuniversitetet i Uppsala)

Address: Box 7021
750 07 UPPSALA
Sweden
Tel. 0 18 171910

Division of Sanitary Engineering, Chalmers University of Technology (Institutionen för VA-Teknik, Chalmers Tekniska Högskola, Göteborg)
Address: 412 96 GÖTEBORG
Sweden
Tel. 0 31 721000

Department of Environmental Engineering, Lund Institute of Technology, University of Lund (Avdelningen för VA-Teknik, Lunds Tekniska Högskola)
Address: Box 118
221 00 LUND
Sweden
Tel. 0 46 107000

TEM FOUNDATION, University of Lund (Stiftelsen TEM)
Address: Box 62
275 00 SJÖBO
Sweden
Tel. 0 416 27300

9.0 Spreading of Information

9.1 Exhibitions

Once every year in May/June, there is a smaller exhibition dealing exclusively with solid wastes in connection with the annual conference of the Swedish association of public cleansing and solid waste management. The place varies from year to year. Hitherto there has been a large exhibition every third year in the city of Jönköping in connection with the conference "ELMIA-Waste".

9.2 Conferences

Conferences take regularly place on the occasion of the above-mentioned exhibitions. The conferences of the Swedish association of public cleansing and solid waste management is normally held in Swedish. From some sessions at the "ELMIA-Waste" conference, simultaneous translation in French, German and English is provided.

9.3 Scientific and technical journals

"RVF-Nytt" ("RVF-News"), edited by RVF, The Swedish Association of Public Cleansing and Solid Waste Management; four issues/year for about 50 pages per issue, completely dealing with solid wastes. First published in 1987.

"Miljö Aktuellt", edited by the National Swedish Environmental Protection Board; ten issues/year for about 20 pages per issue, 15% of which dealing with solid wastes. First published in 1973.

"Miljö i Sverige", edited by Kommunmedia AB; eight issues/year for about 50 pages per issue, 40% of which dealing with solid wastes. First published in 1973.

"Energimagasinet", edited by Teknikförlaget; eight issues/year for about 44 pages per issue, 10% dealing with solid wastes. First published in 1980.

"Stadsbyggnad", edited by the Swedish Association of Municipal Technici; six issues/year for about 66 pages per issue, 15% of which dealing with solid wastes. First published in 1935.

All the above mentioned journals are only published in Swedish.

10.0 References

- "Recycling and handling of solid waste". Governmental Bill 1975:32.
- "Operational Study Solid Waste Treatment/DRAV". The National Swedish Environmental Protection Board and The Swedish Association of Public Cleansing and Solid Waste Management/RVF 1985.
- "Waste Management in Sweden. A National Report". RVF 1984.
- "Waste Treatment in the Future in Sweden—A Combination of Many Methods". RVF 1985.
- "Hazardous Waste Management in Sweden". RVF—Report prepared for ISWA Working Group on Hazardous Wastes 1985.
- "Energy from Waste". The National Swedish Energy Administration and The National Swedish Environmental Protection Board 1986.
- "On Waste Management etc.". Official Government Paper 1986/87:157.
- "Guidelines for Sludge Utilization in Agriculture". The National Swedish Environmental Protection Board 1987.

17

United Kingdom

Surface: 215 631 km^2
Population: 55 836 000 inhabitants
Municipalities:
- Responsible for collection: 423
- Responsible for disposal: 162 (Waste Disposal Authorities—WDAs)

1.0 Structure of Waste Management in the United Kingdom

Waste management systems in the U.K. have been developed to fit with the fundamental principles of E.C.C. directives on the subject, principal among which are a general (framework) directive, 1975 and a directive on toxic and dangerous wastes, 1978.

Legislation governing the management of wastes is developed, drafted and enacted by central government, through the Department of the Environment. The basic roles of central government may be summarized as:

- Development and drafting of legislation;
- Inspection of disposal facilities and operations;
- Incorporating an overseeing role without enforcement powers;
- Research and development (professional and financial support);
- Provision of technical guidance in the form of waste management papers;
- Determination of appeals.

Local government is responsible for applying the control mechanisms defined by law, and for certain operational duties, which apply principally to household wastes. Other authorities which play a significant role, particularly in respect of the disposal of wastes, are planning authorities and water authorities.

1.1 Waste disposal

Waste disposal regulation is the responsibility of the waste disposal authorities, that is the county councils and metropolitan district councils (or statutory authorities) in England and the district councils in Scotland and Wales. The waste disposal authorities are also responsible for disposal operations relating to household wastes.

There is a large private sector waste disposal industry engaged in the collection and disposal of industrial wastes and smaller proportions of commercial and household wastes; the latter under contract to the district councils.

The present system is largely based on the Control of Pollution Act, 1974 and includes:

- A licensing system in which a waste disposal site licence is necessary to operate any waste disposal facility disposing of "controlled" wastes; that is "household, industrial and commercial wastes or any such waste";
- A consignment note system applied to wastes which are dangerous to life, (Special Wastes);
- A duty upon waste disposal authorities to apply the control machinery, that is to issue licences, to monitor and to supervise the licensing and consignment system;
- A duty upon waste disposal authorities to conduct a survey of controlled waste arisings;
- A duty upon waste disposal authorities to prepare long-term plans for the disposal of wastes.

The site licence specifies the types, quantities and proportions of wastes, engineering requirements and the methods of operation.

Consignment notes must be completed by the producer, transporter and disposal of wastes. They include a description of the materials, the form in which they are transported and the place of disposal, which must carry an appropriate site licence.

Landfilling is the predominant method used for the disposal of wastes. This applies to special wastes as it does to household and general industrial wastes, although a significant proportion are subjected to treatments such as incineration and neutralization.

Extensive research and development has taken place, particularly during the last decade, and here again, the most significant developments have been in "landfill science". Considerable research has also taken place in the fields of recycling and the use of waste materials as an energy source. Central government encourages recycling in principle, but does not interfere with market forces.

Although there are specific regulations governing "special wastes" in the U.K., these are integral with the licensing and control system which covers all "controlled" wastes.

1.2 Waste collection

Waste collection is the responsibility of the District Councils. The 1936 Public Health Act contains provisions enabling district authorities to collect household wastes. The use of the power to do so, once undertaken, effectively becomes a statutory duty. There are similar provisions relating to "trade" or "commercial" wastes, the emptying of cesspools and the cleansing and watering of streets. The same legislation empowers local authorities who have undertaken these duties, to make certain requirements on the receiver of the service in connection with the performance of those duties.

Under the Public Health Act 1961, local authorities may provide receptacles for refuse or litter in streets and public places.

Pending legislation includes the imposition of a duty on "collection authorities" (district councils) to collect household wastes and a duty to seek competitive tenders for the collection of household wastes. Traditionally, the collection of household wastes has been carried out by local authorities, but the proposed legislation may increase the small number of areas where collection is already "contracted out" to the private sector.

2.0 Addresses

2.1 ISWA National member

Institute of Wastes Management,
3 Albion Place
Derngate,
Northampton NN1 1UD
England
Tel. 0604 20426 7 8
Secretary-General: R. A. Bispham

ISWA National representative:
D. W. Jackson MBE,
10 Valebrooke,
Tunstall Road,
Sunderland,
Tyne & Wear SR2 7HT
Tel. 091 567 5187
Telex. 537250 ROSE G.

2.2 Government department (Waste Management Functions)

Department of the Environment, (DOE)
Romney House,
43 Marsham Street,
London SW1P 3PY
England
Tel. 01 212 3434

2.3 Private sector

National Association of Waste Disposal Contractors,
26–29 Wheatsheaf House,
Carmelite Street,
London EC4Y 0BN
England
Tel. 01 353 1961
Assistant Director: H. G. Wilson

2.4 Professional association

Institute of Wastes Management

The Institute of Wastes Management was formed over 80 years ago and is the leading professional body in wastes management in the U.K. It's members are drawn from all sectors of the waste industry, including central government, local government, various branches of the private sector (consultants, plant manufacturers, plant and facility operators, contractors), and the academic world.

The aims of the Institute of Wastes Management include:

- "The promotion of scientific, technical and practical aspects of solid and semi-solid wastes management including the collection, treatment, utilization and disposal of domestic, street, commercial, trade and industrial wastes for the safeguarding of the environment."
- There are many ways in which the Institute of Wastes Management pursues this objective and these include the use of fora in the forms of conferences, symposia, research, subject related working groups and study courses. The institute also participates in the consultation processes which precede statutory instruments and in research and development.

There is a general belief within the institute that, there is a need for greater professionalism in wastes management. In pursuance of this, the Institute is currently engaged in an in-depth review that the new structure will include:

- Proficiency tests for operatives;
- A Certificate of competence in wastes management to be held by directors, managers and supervisory staff of waste management organizations;
- Diplomas/Degrees in waste management;
- Election to corporate membership of the Institute of Wastes Management.

2.5 Waste management research

The principal waste management research effort in the U.K. is coordinated and financed by the Department of the Environment (see Section 2.2 above), using the facilities available at government and semi-government institutions, universities, colleges, municipalities and consultants. The current programme for 1987/88 is as follows:

Landwaste Research Programme

(a) Landfill subprogramme

Microbiology, gas and vapours

- Characterization of selected landfill sites;
- Study of methanogens in landfills;
- Laboratory studies on gas production;
- Control processes for chemical and biological stability;
- Field studies on gas pumping collection flaring and vapour;
- Control of gas migration;
- Measurement of methane emission from landfills.

Dynamics of migration and attenuation

- Saturated zone studies at domestic and toxic disposals;
- Mobility and attenuation of microorganic pollutants;
- Liquid disposal into aged-refuse;
- Migration from landfills in Magnesian Limestone;
- Monitor permeable lining beneath Stangate landfill;
- Modelling organic migration beneath Stangate.

Leachate generation and treatment

- Aerobic treatment of ammonia-rich leachate;
- Effectiveness of impermeable liners formed *in situ*;
- Study of drainage problems within landfills;
- Irrigation of leachate to woodland and grass;
- Accelerated leaching by recirculation with treatment.

Reinstatement of landfills

- Mechanical sorting "fines rejects" as a growing medium;
- Composting fines rejects as a soil substitute;
- Utilization of selected waste as a soil substitute;
- Infiltration through clay capping;
- Data on cover, slopes and erosion;
- Waste settlement with various placing methods;
- Effect of leachate on building materials.

(b) *Recovery and sorting—subprogramme*

Mechanical sorting

- Technical evaluation of mechanical separation systems;
- Experimentation of third generation sorting equipment, particularly with front-end technology.

Waste-derived fuel and emissions from waste combustion

- Marketability of WDF including combustion trials;
- Analysis of gaseous emissions from the incineration of waste;
- Technology and control of production and performance to optimize WDF economy on second generation plant;
- Investigation of combustion controls and emission equipment on waste incinerators.

Promoting efficiency in waste management

- Relate household waste arisings to various types of household using survey data and physical analysis;
- Determine component ratios of wastes from various commercial sources.

Recovery and recycling

- Separation for recycling at civic amenity sites;
- Continue development to use rejects from mechanical waste sorting;
- Separation of waste at source and its recycling effects;
- Effect of wheeled-bin collection on recycling;
- Reclamation of growing material from inert wastes;
- Recycling of glass;
- Assessment efficiency/economy of refuse collecting equipment.

(c) *Toxic waste treatment—subprogramme*

- Develop biological ways of breaking down persistent organochlorides and study organic optical sensing for heavy metal ions;
- Study problems in waste management including solidification; immobilization, CCA wood, acid tars and other recalcitrant wastes;
- Survey and analyse the incidence of dioxins and furans in soil for background data and fall-out from waste incineration;
- Biological methods for degrading low-level PCB contamination;
- The modification of clays by organic salts to fix toxic organic wastes;
- Enzymatic methods of reducing the heavy metals content of sludges;
- The use of molten salts to clean up waste oil sludges.

3.0 Spreading of Information

3.1 Scientific and technical journals

The only publication devoted exclusively to the subject is "Wastes Management", the monthly journal of the Institute of Wastes Management (see Sections 2.1 and 2.3).

In addition, the Institute's technical committees produce publications on all aspects of waste management operations. The current list being as follows:

	£
• Task and Finish System of Waste Collection	2.50
• Code of Practice for Site Inspections	2.50
• Advice on Storage and On-Site Treatment of Household, Commercial and Industrial Wastes	4.00
• Transfer Stations	5.50
• Health and Safety in Wastes Management and Allied Services	5.00
• Street Cleansing, Part 1	1.50
• Street Cleansing, Part 2	1.50
• The Analysis of Domestic Waste	2.50
• Basic Landfill Management	1.50
• Winter Maintenance in an Urban Environment	1.50
• Contract Waste Collection—Working Together	4.50
• Technical Publication Binder	2.50
• The Best of Waste Disposal	5.50
• Wheeled Containers—"A Research Study"	5.50
• Register of Consultants	10.00

The Department of the Environment, at four-monthly intervals, publishes a waste management bulletin which appears as a supplement to the Institute's journal and in three other trade publications whose readership includes persons who have an interest in some aspects of waste management.

3.2 Conferences and exhibitions

The Institute of Wastes Management annually organizes the only U.K. conference and exhibition devoted exclusively to this subject. This is held in June each year at various venues and attracts some 800 delegates and 140 exhibitors.

The ninety-first annual conference will be held at Scarborough from June 20–23, 1989. Full details may be obtained from the Secretary-General, Institute of Wastes Management, 3 Albion Place, Northampton NN1 1UD.

3.3 Training courses

In conjunction with regional colleges throughout the U.K., training courses are organized by the Institute of Wastes Management leading to a diploma and corporate membership of the institute. Courses are also organized for junior management level supervisory staff.

4.0 Production and Composition of Solid Wastes

4.1 Domestic, commercial and industrial wastes

Table 1 illustrates the principal waste streams in the U.K. From this, it will be seen that the 423 municipal collection authorities deliver some 18 million tonnes of household and commercial to the 162 municipal waste disposal authorities (WDAs). The WDAs also dispose of a further 14.3 million tonnes delivered direct to their disposal facilities by the general public (Civic Amenity Wastes) and by private sector waste disposal contractors. Of this, 32.3 million tonnes, 4.9 million tonnes is disposed of at private sector facilities, the remaining 27.4 million tonnes being dealt with at municipal facilities.

Table 1 The principal waste streams in the United Kingdom (million tonnes per year)

	WDA disposal		
Collection	WDA sites	Private sites	Private sector disposal
Collection Authorities (18.0)	13.1	4.9	
• Domestic (16.6)			
• Commerce (1.4)			
Civic Amenity Wastes (4.7)	4.7		
Industry and Commerce			
• Commercial (7.8)	7.8		
• Special wastes (1.6)			1.6
• Other hazardous (2.4)			2.4
• General industrial (38.2)			38.2
• Inert wastes (15.0)	1.8		13.2
• Inert wastes, in home use (10.0)			

Note: Excluding mining, agricultural and power station wastes.

Private waste disposal contractors dispose of 38.2 million tonnes of general industrial waste at their landfill facilities, together with 4.9 million tonnes of wastes from WDAs and 13.2 million tonnes of inert wastes, a total of 56.3 million tonnes.

4.2 Other non-controlled wastes

Table 2 illustrates the other principal waste streams in the U.K. The acronym RWA refers to Regional Water Authorities.

Table 2 Other principal waste streams (million tonnes per year)

Collection	Disposal
Agricultural wastes (250.0)	*In situ* or to RWA
British Coal (50.0)	Land Restoration
Minerals Industry (25.0)	Landfill
Power Stations (12.0)	Land Restoration
Total (337.0)	
Miscellaneous Services	
• Cesspits	RWA Treatment Plants
• Gullies	RWA Treatment Plants
• Other Liquids	RWA Treatment Plants

Source: The Future for Rubbish by R. Hawkins and G. Bailey

4.3 Hazardous wastes

In addition to the wastes detailed in Tables 1 and 2, 4 million tonnes of hazardous wastes are disposed of in the U.K. Of these, 1.6 million tonnes are classified as "special wastes" under present legislation. These must be accompanied by a consignment note from point of collection to point of disposal at a facility licensed to deal with such waste, and the WDAs concerned receive a copy of such consignment notes. For landfill disposal at an approved site, the operator must maintain a register of the actual disposal location and, on closure of the landfill, this register must be handed over to the WDA.

Table 3 Treatment streams for hazardous wastes (private sector collection and disposal, millions tonnes per year)

Special waste regulations	
• Landfill	0.75
• Solidification	0.14
• Incineration	0.07
• Chemical treatment	0.30
• Sea disposal	0.28
• Mine shafts	0.06
Other hazardous wastes	
• Landfill	2.40
Total	4.00

Source: Hazardous Wastes Inspectorate 2nd Report (July 1986).

Table 4 Analysis data for a calendar year (percentage by weight)

Constituents	1973	1974	1975	1976	1977	1978	1979	1980
Minus 20 mm Screenings	16.5	16.6	11.8	15.0	10.3	9.0	6.3	7.1
Vegetable and Putrescible	21.6	21.8	23.5	25.8	25.3	21.0	23.0	23.4
Paper	33.0	29.4	30.9	28.9	29.9	35.9	37.3	33.9
Metal	9.5	9.5	9.2	8.0	8.1	8.0	7.2	7.1
Textile	2.6	3.4	3.2	2.2	2.7	2.5	3.5	4.1
Glass	10.9	10.5	12.4	11.3	11.3	11.9	12.0	14.4
Plastic	3.5	4.0	3.8	4.6	4.5	4.5	3.8	4.2
Unclassified	2.4	4.8	5.2	4.2	7.9	7.2	6.9	5.8
Density								
(cwt/y^3)	2.13	2.45	2.17	2.44	2.25	2.17	2.21	2.17
kg/m^3)	142	163	144	162	149	144	147	144

Table 3 illustrates the tonnage passing to each of the six types of disposal facility.

4.4 Composition of municipal solid wastes

The physical analysis of municipal solid wastes is subject to local and seasonal variations but the currently-accepted national analysis is given in Table 4.

5.0 Treatment and Disposal

5.1 Municipal

Of the 32.3 million tonnes of wastes handled by municipal WDA, the most common method of disposal is by landfill, as shown in the following table:

Method	%
Landfill (direct or through transfer station)	88
Incineration	11
Resource Recovery	1
Total	100

Latest 1987 figures for municipal solid waste treatment plants in U.K. are as follows:

Method	No. of plants
Shredding/Pulverization	14
Baling	18
Incineration (with heat recovery)	4
Incineration (without heat recovery)	29
Resource Recovery (waste-derived fuel)	5

5.2 Private sector

Excluding hazardous waste, all private sector tonnage is disposed of by landfill, either directly or through transfer stations.

6.0 Financial Data

The collection and disposal of wastes in the U.K. is now a multimillion pound business. Excluding mining, minerals, power station and agricultural waste, the overall cost of collection and disposal of controlled waste in 1987 was estimated to be as follows:

		£ millions
Municipal		
Collection		488.8
Disposal		298.1
		786.9
Private Sector		
Collection		1006.7
Disposal		465.5
		1463.2
	Total:	2250.1

18

United States of America

Surface: 3 618 770 Square Miles
Population: 227 000 000 inhabitants
Number of cities and rural places: 22 529 of which:

Urban (Cities)
- 6 with more than 1 million population
- 167 with 100 000 to 1 million population
- 2730 with 10 000 to 100 000 population
- 5862 with less than 10 000 population

Rural
- 13 764 with less than 2500 population

1.0 Waste Policy

During the last 20 years, landfills that hold municipal solid waste have decreased from an estimated 18 000 to today's estimated 9000. 25 years ago, land was more plentiful and regulating garbage disposal was not a high priority. During the late 1960s and early 1970, there was an increased public awareness of environmental problems. From the mid-1970s to the present, the public has shifted from environmental awareness to environmental action. The intense public scrutiny of municipal landfills, and the public demand for a cleaner safe environment, have produced several recent changes to the way a municipal landfill conducts business in America.

- In 1986, more than 130 000 new businesses were required to manifest their waste if they generated more than 100 kg/month. This reduced the amount of hazardous waste that landfills could receive. The long-term impact of this law on municipal leachate has not been quantified, but would be positive in nature.
- New regulations will be in place by 1989 that will change the standard for clean drinking water. The number of constituents measured will increase from 25 to more than 83. The constituent level in many cases will be modified downward. They will have a long-term impact on what is considered safe, contaminant-level emissions to groundwater.
- In 1989 the U.S. EPA will be required to enforce a law that bans all land disposal of hazardous waste, unless hazardous waste landfill can be proven to be less risk that other available technology. The stringent

requirements, and high standards of double lining, complex capping systems, leachate detection, ambient air detection, post-closure care, and long-term liability, have generated changes in the design of municipal solid waste landfills.

- For the past several years, EPA has been examining a new testing method for extracting analytical constituents from solid waste/leachate. The proposed test is more rigorous in nature, and could place municipal solid waste incinerator ash into the hazardous waste category.
- Consideration is being given in the proposed regulations governing the disposal of municipal solid waste to banning certain locations, such as fault zones, wetlands, karst terrains, and flood plains, from receiving municipal solid waste. If promulgated at the national level, this could seriously impact the sitting problems many states already are experiencing.

In a November 1987 report to Congress U.S., EPA made the following statement:

"Approximately, 85% of the municipal solid waste generated in 1984 was disposed of in landfills, while only 10% was recycled and only 5% were combusted. Extensive reliance on land disposal for municipal solid waste has resulted in municipal solid waste landfill capacity shortages in some areas of the nation. More than two-thirds (69%) of all municipal solid waste landfills are expected to close within 15 years, 22% are expected to close in 16 to 55 years, 4% are expected to close in 56 to 100 years, and 6% have a life expectancy greater than 100 years. These shortages are becoming critical in densely-populated areas of the country, particularly in the North-east. In addition to the limited use of recycling and waste minimization, other factors that appear to be contributing to the capacity problem include difficulty in siting new disposal or treatment facilities due to public concerns and limited long-term planning by some state and local governments."

Owners and operators of landfills evaluate alternative means of disposing of solid waste when their landfill closes. Based on U.S. EPA's data, the opportunity for change will be high during the next 15 years.

2.0 Addresses

2.1 ISWA National member

Institute for Solid Wastes (ISW)
American Public Works Association (APWA)
Norman Crampton, Secretary, ISW
William S. Forester, Director ISW/ISWA Secretariat
1313 East 60th Street
Chicago, Illinois 60637
Tel. 312 667 2200

- ISW Governor and ISWA Governing Board: John H. Skinner, PhD. Director, Office of Environmental Engineering and Technology Demonstration

  U.S. Environmental Protection Agency
  401 M. Street S.W.
  Washington, D.C. 20460
  Tel. 202 382 2600

- ISW Member and Member of the ISWA Working Group on Sanitary Landfills
  Marcia Williams
  Director, Office of Solid Waste
  U.S. Environmental Protection Agency
  401 M. Street S.W.
  Washington, D.C. 20460
  Tel. 202 382 4627

2.2 Government offices

The lead government office in regulating and developing solid waste programmes at the federal level for disposal of solid waste is:

- U.S. Environmental Protection Agency
  Lee M. Thomas, Administrator
  401 M. Street S.W.
  Washington, D.C. 20460

Other national offices that coordinate with U.S. EPA are:

- U.S. Corp of Engineers
  Lt. General E. R. Heibera III, Commander
  Washington, D.C.

- U.S. Department of Energy
  Donald Walters, Acting Director Energy from M.S.W.
  Department of Energy
  Washington, D.C. 20545

2.3 Technical and professional associations

- Institute for Solid Waste (ISW)
  American Public Works Association (APWA)
  Attention: Mr. Norman Crampton, Secretary ISW
  1313 E. 60th Street
  Chicago, Illinois 60637
  Tel. 312 667 2200

This is an association of municipal engineers, solid waste managers (public and private), consultants and educators that meet at the local, state, and national level. Major committees are disposal, resource recovery, collection and hazardous waste. The Institute of Solid Waste is also the official ISWA member.

- National Solid Waste Management Association (NSWMA)
  Attention: Mr. Eugene J. Wingerter, Executive Director
  1730 Rhode Island Avenue N.W.
  Suite 1000
  Washington, D.C.
  Tel. 202 861 0708

This organization represents the interests of the private solid waste managers at the state and national level. The association has a disposal advisory committee.

- Governmental Refuse Collection and Disposal Association (GRCDA)
  Attention: Mr. Lanier Hickman, Jr., Executive Director
  P.O. Box 7219
  Silver Springs, MD 20910
  Tel. 301 585 2898

This organization is made up primarily of public solid waste officials. It represents their interest at the state and national level and has a separate disposal committee.

- Association of State and Territorial Solid Waste Management Officials (ASTSWMO)
  Attention: Mr. Tom Kennedy, Executive Director
  444 North Capitol Street N.W.
  Washington, D.C. 20001
  Tel. 202 624 5828

This association is made up of state solid waste officials and is very active in providing an advisory role to U.S. EPA on solid waste related issues. This association has a separate disposal committee.

2.4 Main research institutions

- U.S. Environmental Protection Agency
  Attention: Solid Waste Research Division (OSW)
  26 W. St. Clair Street
  Cincinnati, OH 452368

- United States Corporation of Engineers
  Construction Engineering Research Laboratory (CERL)
  Attention: Environmental Division
  P.O. Box 4005
  Champaign, IL 61820

While there are other governmental offices that provide research in solid waste, U.S. EPA and CERL, have provided significant contributions during recent years. Most of the solid waste research in the U.S. is done through a combination of universities, consultants, public agencies and private corporations. Most of the pure research is funded by the federal government. Professional associations (Section 2.3) are beginning to fund small research projects. Some states have explored a special tax on landfills to help fund needed solid waste research.

3.0 Spreading of Information

3.1 Scientific and technical journals

The following is a list of 14 journals and publications that contribute to body of knowledge of solid waste and are published in the U.S.

- Journal of Environmental Engineers, edited by:
  American Society of Civil Engineers (ASCE)
  345 East 47th Street
  New York, NY 1017–2398
  Publishes 120 pages monthly, 4% directly addresses solid waste.

- Power Engineering, edited by:
  Ban Enterprises
  1620 McQueen Street
  Barrington, IL 60010
  Published 60 pages monthly, 3% directly addressed solid waste resource recovery facilities.

- World Wastes, edited by:
  Communication Channels
  6255 Barfield Road
  Atlanta, GA 30328
  Publishes 60 pages monthly with an extra issue in December. All articles deal directly with the solid waste industry.

- American City and County, edited by:
  Communication Channels
  6255 Barfield Road
  Atlanta, GA 30328
  Published 90 pages monthly, 5% deals directly with solid waste issues facing cities and counties.
- The Journal of Resource Management and Technology, edited by:
  Iraj Zandi, Professor of Resource Recovery and Management
  220 S. 33rd Street
  Towne Building/D3
  University of Pennsylvania
  Philadelphia, PA 19104
  Publishes 120 pages quarterly. All paper deals with the resource recovery aspects of solid waste.
- Solid Waste Proview, edited by:
  Institute of Solid Waste (ISW)
  American Public Works Association
  1313 East 60th Street
  Chicago, IL 60637
  Publishes 8 pages quarterly. Each publication features one significant feature article, two interviews, a professional profile of an institute member, and an annotated bibliography of 18–20 articles.
- Operations Forum, edited by:
  The Water Pollution Control Federation
  601 Wythe Street
  Alexsandria, VA 22314–1995
  Published 30 pages monthly. 20% of the magazine deals with sludge processing and disposal techniques.
- Public Works, edited by:
  Public Works Journal Corporation
  Office of Publication
  200 S. Broad Street
  Ridgewood, NJ 07451
  Publishes 100 pages monthly with an extra publication in April each year. 15% of the articles feature solid waste.
- APWA Reporter, edited by:
  American Public Works Association
  Rodney Fleming, Editor
  American Public Works Association
  1313 E. 60th Street
  Chicago, IL 60637
  Publishes 40 pages monthly. 17% of the articles feature solid waste.

- Biocycle, edited by:
  J. G. Press Inc.
  Box 351
  18 S. Seventh
  Emmaus, PA 18045
  Publishes 60 pages monthly with combined issues in May–June and November–December. Primarily addresses sewage sludge utilization.

- American Water Works Journal, edited by:
  American Water Works Association
  6666 W. Quincey Avenue
  Denver, CO 80235
  Publishes 170 pages monthly describes major issues surrounding drinking water quality. A good reference for "how clean is clean".

- Journal of Water Pollution Control Federation, edited by:
  Water Pollution Control Federation
  601 Wythe Street
  Alexsandria, VA 22314–1994
  Published 100 pages monthly, articles center around sewage treatment processes and sludge technologies.

- Civil Engineering, edited by:
  American Society of Civil Engineers
  345 East 47th Street
  New York, NY 10017–2398
  Publishes 110 pages monthly, 2% addresses solid waste.

- City Currents, edited by:
  National Resource Recovery Association
  United States Conference of Mayors
  1020 Eye Street N.W.
  Washington, DC 20006
  Publishes 30 pages bimonthly. Concentrates on status of resource recovery facilities in the U.S.

Other publications exist in the form of weekly or monthly newsletters that are used to keep solid waste managers abreast of key solid waste issues; however, these have limited circulation and can cost up to $400/year.

3.2 Exhibitions

There are three major exhibitions that display large amounts of solid waste handling and treatment equipment. They are:

- International Public Works Congress and Equipment Show, sponsored by:
  American Public Works Association
  1313 E. 60th Street
  Chicago, IL 60637

The 1987 Congress attracted 9513 registrants and featured 466 exhibitors. The 1988 Congress will be held in Toronto, Canada.

- Waste Exposition, sponsored by:
  National Solid Waste Management Association
  1730 Rhode Island Avenue N.W.
  Suite 100
  Washington, D.C. 20036

The 1987 Waste Exposition, was held in Dallas, Texas and attracted 3400 registrants and 312 exhibitors. The 1988 Waste Exposition will be held in Washington D.C.

- International Solid Waste Exposition, sponsored by:
  GRCDA
  P.O. Box 2219
  Silver Springs, MD 20910

The 1987 Exposition was held in St. Paul, MN and had 1000 registrants with 116 exhibitors. The 1988 Exposition will be held in Baltimore, MD.

Small conferences are held throughout the U.S. by chapters of the larger organizations.

3.3 Conferences

Conferences take place at each of the major exhibitions (see Section 3.2). Speciality conferences are held throughout the U.S. during each year by APWA, NSWMA and GRCDA. Monthly and annual meetings are held by state chapters of these organizations (Section 2.3). The national solid waste exhibitions each have three to four days of technical sessions and tours of local solid waste facilities.

Other significant conferences held annually in the United States are:

- Symposium on Solid Waste Testing and Quality Assurance, sponsored by:
  U.S. Environmental Protection Agency
  401 M. Street S.W.
  Washington, D.C. 20460

The four-day symposium features recent developments in testing solid waste and is held annually in Washington, D.C. Over 800 environmental engineers and scientists from industry and government attended the 1987 symposium.

- HazMat International is sponsored by:
  Chemical Equipment Magazine
  Tower Conference Management Co.
  800 Roosevelt Road Building E-Suite 408
  Glen Ellyn, IL 60139–5835

The three conferences are held annually in Atlantic City, N.J. The 1987 conference had 5700 registrants and 325 exhibitors.

3.4 University courses

Solid waste courses are generally held in major universities and conducted through the civil engineering departments. In depth training for scientists, engineers and solid waste managers is provided by many universities through their continuing education or extention service programmes.

4.0 Production and Composition of Solid Wastes

Solid Waste is regulated under substitute D of the Resource Conservation and Recovery Act (RCRA). The definition of solid waste under the act is:

. . . any garbage, refuse, sludge from a waste treatment plant, water supply treatment plant, or air pollution control facility and other discarded material, including solid, liquid, semisolid, or contained gaseous material resulting from industrial, commercial, mining, and agricultural operations, and from community activities, but does not include solid or dissolved materials in domestic sewage, or solid or dissolved materials in irrigation return flows or industrial discharges which are point sources subject to permits under Section 402 of the Federal Water Pollution Control Act, as amended (86 Stat. 880), or source, special nuclear, or byproduct material as defined by the Atomic Energy Act of 1954, as amended.

In accordance with the above-mentioned definition, the following categories of Subtitle D wastes have been identified:

- Municipal solid waste
- Household hazardous waste
- Municipal sludge
- Municipal waste combustion ash
- Infectious waste

- Waste tyres
- Industrial nonhazardous waste
- Very small quantity generator hazardous waste (<100 kg/month)
- Construction and demolition waste
- Agricultural waste
- Oil and gas waste
- Mining waste

Each of the categories are quantified in a 1987 U.S. EPA report to Congress (Table 1).

Table 1 Estimated annual generation rate for different categories of wastes

Waste category	Estimated annual generation rate (million tonnes)
Industrial nonhazardous waste	7600[a]
Oil and gas waste	
• drilling waste	129–871[d]
• produced water	1966–2738[e]
Mining waste	1400
Municipal solid waste	148
• household hazardous waste	0.002–0.5
Construction and demolition waste	31.5
Municipal sludge	
• wastewater treatment	6.9
• water treatment	3.5
Municipal waste combustion ash	4.5
Very Small Quantity[b] Generator Hazardous waste (<100 kg/mo)	0.2
Waste tyres	240 million tyres
Infectious waste	Unknown
Agricultural waste	Unknown
Approximate total	11 300

[a] Not including industrial waste that is recycled or disposed of offsite.

[b] Small quantity generators (100–1000 kg/mo waste) have been regulated under RCRA, Subtitle C, since October 1986. Before this, approximately 830 000 t of small-quantity generator wastes were disposed of in Subtitle D facilities every year.

[c] Waste category is the subject of a separate report to Congress.

[d] Converted to tonnes from barrels: 42 gallon = 1 barrel, 17 lb/gallon.

[e] Converted to tonnes from barrels: 42 gallon = 1 barrel, 8 lb/gallon.

[f] No total annual quantities were available for infectious waste, but it is estimated that 5–9 pounds are produced/patient/day.

Table 2 gives an estimated breakdown of the 133 million tonnes of municipal solid waste, which is a mixture of household, institutional, commercial, municipal and industrial solid wastes.

Table 2 Estimated breakdown of municipal solid waste production

Material	T/year (million)	Percentage (%)
Paper and Cardboard	49.4	37.1
Glass	12.9	9.7
Metals	12.8	9.6
Plastics	9.6	7.2
Rubber and leather	3.3	2.5
Textiles	2.8	2.1
Wood	5.1	3.8
Food Waste	10.8	8.1
Yard Waste	23.8	17.9
Other non-food Product Waste	0.1	0.1
Miscellaneous Inorganic Wastes	2.4	1.8
Total	133.0	100.0

The national average generation rate for municipal solid waste is 3.2 lb/cap/calendar day. This figure will vary with location. Areas with high vegetation and construction activity have reported over 6 lb/cap/calendar day.

5.0 Municipal Solid Waste (MSW) Collection

Precise data has not been developed to show the disposal method used by individual homes or commercial business; however, a general discussion of each method in probable order of usage follows:

Residential

- Bags (paper and plastic). The individual resident normally transfers the refuse to the collection plant by either paper or plastic bags. Many cities offer the bags for sale or give them to the resident to promote cleaner or more efficient collection systems. Engineers designing municipal waste landfills are becoming more cognizant of the non-biodegradable nature of the plastic bags.

- Cans. Normally 20 gallon garbage cans are placed or rolled out at the point of collection. Because of the requirements of can with double handles and requirements that they be returned to the point of collection after the refuse is deposited in the garbage truck, along with the unsightly nature of empty cans left in a front yard after the pick-up, many cities require bags only.
- Carts. Carts are 80–120 gallon molded plastic containers that have wheels. The carts normally have built-in lids and are rolled from the residence to the point of collection on the designated pick-up day. The carts have the additional advantage of either being picked-up automatically with a side loader or semi-automatically with a rear loader; thereby limiting the amount of injuries to the refuse collection work force.
- $1\frac{1}{2}$–8 y^3 Containers (plastic or metal). There are large multi-family containers that are collected with automated collection equipment. This system is predominately used in apartment complexes or in cities that have sufficient alleys to accommodate the containers. Trends in the last 20 years have been increases in urban populations and increases in the number of apartment units.
- Containers larger than 8 to 40 y^3. Some apartment units have large stationary roll-off compactors in which the refuse is deposited and collected.

Commercial

- $1\frac{1}{2}$–8 y^3 Containers (plastic or metal). This is currently the predominate way to remove commercial refuse. The containers are compatible with apartment collection systems and provide for boxes and crates associated with commercial waste.
- Containers larger than 40 y^3. This method of disposal is increasing in popularity as more and more large shopping centers continue to be developed. Large stationary compactors are used in grocery stores and shopping centers to curtail pilfering. Roll-off containers are predominately used for construction debris.
- Bags, Cans and Carts. Commercial businesses still use this system where small quantities of refuse are generated and on built-up areas where the larger containers are not practical.

Other collection information

- Bulky Object Collection. Many cities provide separate periodic collection of trimmed brush and bulky objects (white goods) from the household. This required special equipment (Roto-booms and trucks with drag boards), and in some cases allows the cities to operate brush burn sites to conserve landfill space. The burn sites most often utilize our protection devices to remove particulate matters from the air emissions.

- Dead Animal Collection. Cities provide specially equipped vehicles to collect deal animals. The animals in some cases are incinerated or handled with immediate cover at landfill sites.
- Household Hazardous Waste Collection Programmes. Approximately 35 cities have established special collection programmes for household hazardous wastes. This waste is normally accepted by trained specialists and taken to a hazardous was disposal site.

6.0 Municipal Solid Waste Treatment and Disposal

Landfilling is the most common way of treating solid waste in America (85%), recycling is second (10%), and incineration with recovery is third (5%).

Table 3 Waste processing techniques utilized in a 1986 survey by the American Public Works Association

Technique	% Used
Source separation	40
Shredding	23
Baling	10
Tyre shredding	10
Recycling	7
Incineration	5

Other techniques that currently are utilized at landfills to conserve space, but to a lesser extent are:

- The use of foams as a daily cover for vector control in lieu of 6 inches of soil.
- Surcharging deep fills to increase settlement and increase capacity. (Some day fills of 70–100 feet have reported 20–25% volume reduction by this method.)
- Curbside recycling throughout some communities has increased landfill capacity by an estimated 5%.

- The primary way most landfills conserve space is through the use of special landfills compactors that are designed to shred the waste in place. Landfill operators that use compactors in lieu of conventional track equipment realize volume reductions up to 25%.

Perhaps the most remarkable change in recent years is not in the processing but in the design of landfills. These facilities are becoming increasingly difficult to site in U.S. For that reason engineers have designed landfills to maximum allowable heights. The Fresh Kills Landfill in New York City, NY was projected to be the highest point on the Eastern seaboard of over 500 feet above sea level. The tremendous cost to haul waste to a more distant site makes the land more valuable as a landfill than for any commercial usage. Faced with siting difficulties, it is a common practice for many cities to increase the depth of the landfills.

6.1 Incineration

Incineration and the production of refuse derived fuel from MSW are still emerging technologies in U.S. As opportunities are available to communities through landfill depletion or special energy needs, resource recovery has become increasingly popular.

Table 4 Resource recover status (U.S.A. 1987)

	1–300 TPD	301–800 TPD	801 TPD and above	Total
Operating or in shakedown (84)				
Mass burn	45	8	12	65
RDF	3	7	6	16
Other	1	—	2	3
Under construction (42)				
Mass burn	6	13	12	31
RDF	1	2	8	11
Advanced planning (15)				
Mass burn	4	7	1	12
RDF	1	—	2	3
Total	61	37	43	141

Landfill Gas Recovery Projects
Operating: 35
Under construction: 2
On hold or temporarily closed: 6
Canada: 1

The results of an October, 1987 Survey by the National Resource Recovery Association lists 141 resource recover facilities in U.S. that are operating, in shakedown, under construction, or in advanced planning (construction expected within several months). Six listed facilities are closed; these are either shut down temporarily or expected to be replaced. The following table gives a breakdown of the facilities by status, size and technology.

The report also lists 126 communities with waster-to-energy projects in earlier planning stages.

Modular construction has allowed many of the smaller plants to select mass burn as opposed to refuse-derived fuel. From Table 4, it appears that refuse derived fuel becomes more competitive with mass burning in the larger plants. Smaller cities have financed resource recovery plants when they have found industries that can use the steam in their manufacturing process. In areas of the country that depend on sole source aquifers, environmental standards for landfills are very strict and resource recovery becomes more financially competitive. Several major concerns are facing the industry at this time.

- Environmental concerns and determination that will be made by USEPA concerning stack emissions and incinerator ash. Should the heavy metal content of MSW ash produce a hazardous leachate, then the expense of disposing of the ash may eliminate incineration as an economic alternative.
- The transfer of some European technologies has failed to take into account the higher percentage of glass in the U.S.A. waste stream. Molten glass liquifies in the secondary burn chambers, then solidifies creating a blockage of the air passage from the secondary chamber to the heat recovery unit.

Power Engineering published in March, 1987, a comprehensive article "Energy from Waste". The article accurately depicts several important tables that reflect the composition of MSW, comparative analysis of combustion technologies and typical costs for a modular plant.

Table 5 How composition of RDF compares with unprocessed MSW

Parameter	Raw refuse	RDF
HHV, Btu/lg	4500	6349
Ash (as fired), %	26	9.5
Ash (dry basis), %	36	12.3
Moisture, %	23	23
Sulfur, %	0.42	0.27
Nitrogen, %	0.71	0.57
Carbon, %	36	41

Table 6 Combustion technologies rated best (1) to worst (4)

Parameter	Rotary kiln	Spreader stoker	Bubbling bed	Cir bed
Experience	2	1	3	4
Capital cost, $[a]	4	1	2	3
	(N/A)	(8050)	(8503)	(11 520)
Reliability, availability	3	1	3	2
Operating, maintenance costs[b]	4	1	3	2
Load response, turndown	2	1	4	3
Range of coals	N/A	3	2	1
Coal preparation	N/A	2	3	1
Auxiliary power (hp)	2	1	3	2
Sulfur removal	4	4	2	1
NO_x emissions	4	3	2	1
Sophistication of controls	2	1	4	3
Visual operation	1	4	3	2
Fouling possibility	3	2	4	1
Slaggin possibility	3	4	1	2
Efficiency	4	2	3	1
Explosions	2	1	3	4
Unburned combustibles	2	3	4	1
Fuels in combination	1	3	4	2

[a] Costs solicited from different manufacturers, based on a 200 000 lb/hr (coal, 800 psig/850F system).
[b] Costs would vary based on utilization of waste fuels.

Table 7 Typical costs for a modular plant, $1000

	Case A[a]	Case B	Case C
Capital costs			
Plant construction	1648	3393	8186
Air pollution control	0	400	1620
Annual capital recovery	154.4	355.3	919.6
OAM costs			
Labor	40	200	400
Auxiliary fuel (oil)	8.5	45.9	122.5
Electricity	4.6	24.9	66.5
Maintenance, repair	36.2	196.9	525
Air pollution control	0	12.6	63
Fuel savings (oil)	117.6	516.8	1575

[a] Plant A burns 2412 t of refuse annually; B, 13 125 t; C, 35 000 t.

6.2 Material recovery and compost

Of the facilities shown in Table 3, 36 recover a resource other than steam or electricity.

Table 8 Material recovery chart

Product	Number of plants
Ferrous metal	36
Aluminum	8
Compost	2
Glass	3

Raw materials markets have not allowed an abundance of MSW recovery projects to florish in U.S. While environmental groups would prefer larger percentages of our MSW stream to be recovered, established raw material markets, high recovery costs and no tax increase budgets have not allowed municipal decision-makers to commit to long term recycling or recovery projects.

Two cities—Austin, Texas and Fresno, California—have had agressive and popular curbside recycling programs for many years. Several city councils here asked administrators to plan for long-term recycling of waste as opposed to mass burning. It was estimated in Austin, TX that a maximum curbside recycling program could reduce landfill volume requirement by 5%.

By contrast composting of municipal sewage sludge is practiced widely in U.S. In-vessel composting, beneficial land application, static pile composting and pelletizing are all technologies that are currently being applied to allow the utilization of domestic wastewater sludge as a resource.

6.3 Sanitary landfill

A typical municipal solid waste landfill incorporates, as a minimum, the following environmental control features, in its design:

Control Feature	Standard
Flood Protection	To accommodate a 100 year flood frequency
Periphery Groundwater Monitoring Wells	1 year up-gradient well and 2 down-gradient wells
Groundwater Protection	Minimum 42 inches of clay with a periability of 1 × 10 cm/SCC, or a synthetic liner.

Erosion Control	2–6% final Grade with vegetation cover
Vector Control	Daily cover of 6 inches earth with 2 feet of final cover.

Special design consideration must also be given to methane migration, leachate quality and generation, final cover, blowing paper, daily operations, fill sequencing, and post-closure care.

In a draft report to Congress by USEPA, a comprehensive survey was made of actual environmental features incorporated in MSW landfills in the USA. The information in Table 9 reflects a portion of the survey.

Table 9 Types of release prevention methods

Management method	Landfills surveyed, 6000
Synthetic liners	73
Natural liners (e.g., clay), including slurry walls	3409
Leachate collection systems	732
Runon/runoff controls	4016
Methane controls (vents, recovery)	123
Leachate treatment (except leachate recirculation)	641
Restrictions on receipt of liquid wastes (e.g., bulk liquid restrictions)	No data

Tables 10 and 11 are taken from the 1987 USEPA report to Congress and reflect the number and acreage of MSW landfills in the USA.

Table 10 Number of landfills by ownership category

		Ownership category					
Landfill type	Response rate	Owned by state government	Owned by local government	Owned by federal government	Privately owned	Other	Total
Municipal waste	96%	126 (1.4%)	6908 (77.9%)	348 (3.9%)	1482 (16.7%)	8 (0.1%)	8872 (100%)
	97%	49[a] (0.8%)	3343[a] (57.1%)	193[a] (3.3%)	802[a] (13.7%)	1465[a] (25%)	5853[a] (100%)

[a] Reflects a total survey population of 6600.

Table 11 Landfills by acreage category

Landfill type	Response rate	Average category		
		Less than 10	10 to 100	More than 100
Municipal waste	75%	2944 (42.3%)	3572 (51.3%)	449 (6.4%)

Table 12 Violations of state standards at municipal solid waste landfills in 1984

Medium of concern	Violations detected[a]	Facilities with monitoring	
		Number	Percentage (%)
Ground water	586	2331	25
Surface water	660	1100	12
Air	845	358	4
Methane (subsurface gas)	180	427	5

[a] Violation data are based on the 9284 municipal solid waste landfills identified by the 1984 State Census.

Table 12 from the same report reflects the number of violations reports to USEPA by state surveys.

Based on USEPA's data presented in Table 9 and Table 12 it can be determined that good design practices are not being followed when constructing many of the municipal solid waste landfills in U.S.

Increased control of incoming contaminants (very small quantities of hazardous waste), attention to ground water protection and air pollution control will be encouraged by many states as a direct result of the 1987 report to congress.

While ground water contaminant levels are already well established by the clean water act, ambient air contaminant levels surrounding a landfill site in US need to be well defined.

6.4 Addresses of recent installations

Table 13 is a description of 14 new resource recovery facilities that have began operations in US in the last two years.

Table 13 Description of new resources recovery facilities in the USA

Location and major participants	Processes	Products and uses	Capacity (t/day)	Capital costs ($ millions)	Status	Source
Bellingham Thermal reduction Inc. (designer, owner & operator)	Mass burning	Electricity for sale to Puget Sound Power & Light	100	6.5	Operational since 11/86	James R. Anderson President Thermal Reduction, Inc. 1524 Slater Road Bellingham, Washington, 98226
San Marcos (North County Recycling and Energy Recovery Center) San Diego County; North County Resource Recovery Associates (joint venture of Brown & Root and Thermo-Electron) (owner/operator)	Separation, shredding, combustion of RDF, electricity generation	Aluminum, ferrous metals, cardboard, film plastic, compost; electricity for sale to San Diego Gas & Electric Co.	1672	120	Under construction; startup expected in 1989	Sharon Reid Solid Waste Programme Manager County of San Diego Dept. of Public Works 5555 Overland Avenue, 0383 San Diego, CA 92123
Bridgeport Conn. Resources Recovery Authority; Bridgeport and other cities; Wheelabrator Environmental Systems, Inc. (owner/operator)	Mass burning in water-wall combustion units; electricity generation	Electricity for sale to United Illuminating Co.	2250	200	Under construction; operations expected in May 1988	Kevin Stickney Wheelabrator Environmental System, Inc. Liberty Lane Hampton, NH 03842

Hartford Connecticut Resources Recovery Authority (sponsor/owner); Combustion Engineering (designer/builder); Metropolitan District, Conn. Light & Power and Combustion Engineering (operators)	Firing to RDF to produce steam	Steam for sale to Connecticut Light & Power Co. for use in electricity generation	D-2000	154	Construction underway since 5/85; start up began in 9/87, with commercial operations to begin in 6/88	Thomas F. Lenehan Project Manager Connecticut Resources Recovery Authority 179 Allyn Street Hartford, CT 06103
Tulsa City; Ogden Martin System of Tulsa, Inc. (owner/operator); Kiewit Industrial Corp. (general contractor)	Mass burning, generation of steam and electricity	Steam for sale to Sun Refining & Marketing Co.; electricity for sale to Public Service Co. of Okla.	750; expanding to 1125	76.0	Operational since 3/86; expansion to 1125 TPD under construction with operation expected in 10/87	Herbert E. Van Fleet Director Dept. of Refuse City of Tulsa 470 W. 23rd Street Tulsa, Okla. 74107
Marion County County; Ogden Martin Systems, Inc. (owner/operator/designer)	Mass burning in waterwall furnaces	Electricity (11 Mw) for sale to Portland General Electric Co.	550	47.5	Operational since 4/86; first waste-to-energy plant in U.S. with dry scrubbers and baghouses	Jim Spears Director of Solid Waste Room 300 220 High N.E. Salem, Ore. 97301

Continued

Table 13 *Continued*

Location and major participants	Processes	Products and uses	Capacity (t/day)	Capital costs ($ millions)	Status	Source
City of Commerce Commerce Refuse-to-Energy Authority (owner); Los Angeles County Sanitation Districts (operator); Henningson, Durham & Richardson (designer)	Mass burning in waterwall incinerator	Electricity for sale to Southern California Edison	300	35.0	Operational since 5/87	Mike Selna Project Manager Commerce Refuse-to-Energy Authority P.O. Box 4998 Whittier, CA 90607
Tampa City (owner); Waste Management, Inc. (design/construct/operate)	Mass burning	Electricity to be sold to Tampa Electric Co.; scrap metals	1000	70	Operational since 9/85	Nancy McCann McKay Bay Refuse-to-Energy Project City Hall Plaza 5 North Tampa, Fla. 33602
West Palm Beach S.W. Authority of Palm Beach County (owner); Babcock & Wilcox/National Ecology, Inc. (operator); Palm Beach Energy Assoc. (joint venture of B&W and Bechtel) (designer)	Refuse processing to RDF; ferrous and aluminum recovery; burning RDF; electricity generation	Electricity for sale to Fla. Power & Light Co.; aluminum, ferrous metals	2000	185	Groundbreaking expected in 12/87, with startup in 12/89	Timothy F. Hunt, Jr. Executive Director Solid Waste Authority of Palm Beach County 5114 Okeechobee Blvd. Suite 2C West Palm Beach, FL 33417

Savannah City; Katy-Seghers, Inc. (owner/ operator/designer)	Mass burning	Steam for industrial use for sale to Kemira, Inc.	500	35	Operational since 4/87	Gene Prevatt Resource Management Administrator P.O. Box 1027 Savannah, Ga. 31402
Dutchess County Dutchess County Resource Recovery Authority (owner); Pennsylvania Engineering Corp. (design/build/ operate)	Mass burning in Westinghouse O'Connor rotary combustor for generation of steam and electricity; ferrous metals recovery	Steam for sale to IBM Corp.; electricity to utility; ferrous metals	400	30	Project in preacceptance testing stage; operation expected 12/87	C. Scott Daniels Commissioner of Solid Waste Management 22 Market Street Poughkeepsie, NY 12601
Rochester Olmstead County (owner & operator); HDR Techserv, Inc. (designer)	Mass burning in refractory-lines waterwall	Electricity (4.2 MW) and steam for district heating	200	25	Operational since 4/87	Charles R. Michael Project Manager Olmstead Waste-to-Energy Facility 2122 Campus Dr., S.E. Rochester, MN 55904
Newport Ramsey & Washington Counties; Northern States Power (owner, operator, builder)	RDF producing and burning for electricity production	Electricity, ferrous metals, aluminum	1000	31	Operational since 7/87	Patrick Storey Project Manager 1811 Weir Dr. Woodbury, MN 55125

Continued

Table 13 *Continued*

Location and major participants	Processes	Products and uses	Capacity (t/day)	Capital costs ($ millions)	Status	Source
Hennepin County County; Hennepin Energy Resource Co. (Blount Energy Resource Co., gen. partner) (owner/ designer/builder/ operator)	Mass burning	Electricity	1200	70	Construction began in May 1987 with startup in 1989	Warren Porter Solid Waste Division Manager Hennepin County Department of Environment & Energy 822 S. 3rd Street Suite 300 Minneapolis, MN 55415

7.0 Industrial Wastes

It was found that the waste streams of 22 major industries accounted for more than 99% of the nonhazardous industrial waste. A breakdown is provided in Table 14.

A summary of the data available, potential hazard and prevalent waste management method is reflected in Table 15.

Table 14 USEPA Report to congress—industrial waste

Industry	Waste quantity (dry tonnes)	% of total
Industrial organic chemicals	97 354 000	24.8
Primary iron and steel manufacturing and ferrous foundries	60 679 000	15.5
Fertilizer and other agricultural cultural chemicals	59 037 000	15.0
Electric power generation	55 878 000	14.2
Plastics and resins manufacturing	44 991 700	11.5
Industrial inorganic chemicals	26 191 800	6.7
Stone, clay, glass, and concrete products	18 600 000	4.7
Pulp and paper industry	8 627 000	2.2
Primary non-ferrous metals manufacturing and non-ferrous foundaries	6 575 000	1.7
Food and kindred products	6 361 500	1.6
Water treatment	4 960 000	1.3
Petroleum refining industry	1 276 400	0.3
Rubber and miscellaneous plastic	542 600	0.1
Transportation equipment	520 000	0.13
Fabricated metal products	300 000	0.08
Pharmaceutical preparations	256 900	0.07
Machinery, except electrical	193 000	0.05
Lumber and wood, and furniture and fixtures	122 700	0.03
Textile manufacturing	45 000	0.01
Soaps; other detergents; polishing, and sanitation goods	31 300	0.01
Leather and leather products	24 600	0.01
Electrical machinery and electronic	10 400	0.01
Total	392 579 900	

Table 15 Qualitative analysis of industrial nonhazardous waste data

Industry	Data availability[a]	Relative levels of heavy metals or organics in wastes	Prevalent waste management methods
Electrical machinery and electronic components	POOR: The descriptions of waste types are incomplete (Year, 1977)	HIGH: Wastewater treatment sludges, oils, and paint wastes have potential to release heavy metals and organics. No specific analytical data are available. Since this industry generates considerable quantities of hazardous waste, some small-quantity generators may dispose hazardous wastes in on-site, land-based facilities.	General trend indicates off-site landfill disposal, based on 1977 data. Large quantities of non-hazardous wastewaters may be managed in on-site surface impoundments.
Electric power generation	GOOD: Detailed descriptions of waste types and quantities are available. Waste management data are fairly good. (Year—1983)	MODERATE: This waste has a potential to reduce pH levels and release metals. Organics, such as naphthalenes and benzofluorenes, also may be released. Toxicity depends on the source of coal or oil being burned.	General trend is on-site disposal in clay-lined surface impoundments and landfills. Some of these facilities are synthetic-lined and have ground water monitoring.

Fabricated metal products	POOR: Waste type and quantity data are almost completely nonexistent. Some management data are available. (Years—1976, 1979, and 1983)	HIGH: Wastewater treatment sludges, oils, and paint wastes have potential to release heavy metals and organics. No specific analytical data are available. Since this industry generates considerable quantities of hazardous wastes, some small-quantity generators may dispose of hazardous wastes in on-site, land-based facilities.	Data from 1976 indicate that 20–30% of wastes are managed on site in landfills and lagoons.
Fertilizer and Other Agricultural Chemicals	MODERATE: Waste quantity and management data are very good for pesticide formulation and manufacturing, but are poor for some segments of fertilizer manufacturing. Waste types are fairly well-defined for fertilizer and detailed analyses are available for pesticides. (Years—1980 and 1983)	HIGH: Waste gypsum piles may cause local pH and metals contamination problems. Pesticide wastes may release organics and heavy metals.	Waste gypsum is stored in unlined piles. Large quantities of waste waters are stored or treated in surface impoundments.

Continued

Table 15 *Continued*

Industry	Data availability[a]	Relative levels of heavy metals or organics in wastes	Prevalent waste management methods
Food and Kindred Products	GOOD: Waste types and quantities are well defined and waste management methods are fairly well described. (Year—1980)	LOW: Most food industry wastes are biodegradable, but may cause taste and odor problems.	Off-site landfills and land application are used extensively, with some onsite land disposal.
Industrial inorganic chemicals industry	MODERATE: Data on waste quantities and amounts are good, but there are very little analytical data. (Year—1979)	HIGH: Most nonhazardous wastes from this industry do not appear to contain heavy metals, but there are insufficient analytical data on these wastes. Since this industry generates considerable quantities of hazardous wastes, some small-quantity generators may dispose of hazardous wastes in on-site, land-based facilities.	On-site landfills and surface impoundments are used for most wastes. Design data on these facilities are not available.

Industrial organic chemicals	VERY GOOD: Detailed information is available on all data areas except the design features of the waste management facilities. (Years—1981 and 1982)	HIGH: Many of the waste streams in this industry contain high levels of extremely toxic organic chemicals. Since this industry generates considerable quantities of hazardous wastes, some small-quantity generators may dispose of hazardous wastes in on-site, land-based facilities.	Most land-based disposal is performed at off-site facilities; however, approximately 34% of waste water and sludges are treated in on-site impoundments prior to discharge.
Leather and leather tanning	GOOD: Waste types and quantities are well described and general management methods are known. (Year—1975)	MODERATE: These wastes generally contain chromium, but it is generally in the +3-valence state.	Off-site landfills are used most commonly, and approximately 90% of all wastes are sent to off-site facilities. Ten % of the wastes are managed in on-site surface impoundments and landfills.
Lumber and wood products and furniture and fixtures	MODERATE: The waste types in this industry are described, but there are no dependable data on quantities of analytical results. (Year—1980)	MODERATE: Most of the wastes (380 million MT/year) from this industry are composed of wood dust, chips, shavings, and other rejects, and most of these wastes are burned or reused. However, the ash from burning these wastes is generated in very high quantities and is high in pH.	There are no data on land disposal of wastes from this industry.

Continued

Table 15 *Continued*

Industry	Data availability[a]	Relative levels of heavy metals or organics in wastes	Prevalent waste management methods
Machinery except electrical	POOR: The descriptions of waste types are incomplete. (Year—1977)	HIGH: Waste waste treatment sludges, oils, and paint wastes have potential to release heavy metals and organics. No specific analytical data are available. Since this industry generates considerable quantities of hazardous waste, some small-quantity generators may dispose of hazardous wastes in on-site, land-based facilities.	Data from 1977 indicate that 90% of these wastes are managed off site and that 70% of the total waste stream from this industry are land disposed. Ten percent of these wastes are managed on site; however, the management methods are not known.
Pulp and paper industry	GOOD: The quantities and types of wastes from this industry are well described, and management methods are known for each waste type. Some data are available on waste management facility designs. (Year—1977)	MODERATE: Organic pollutants from wood fibers may be significant. Also, coal and bark ash may contain metals. Sulfates and metals are high in some pulping wastes.	Approximately 72% of all wastes are managed in on-site landfill facilities. On-site surface impoundments account for 7% of industry wastes; about 10% of pulp and paper wastes are managed in on-site incinerators.

Petroleum refining industry	VERY GOOD: All data needs were available except typical designs of waste management facilities. (Year—1977)	HIGH: These wastes generally contain high levels of sulfides, ammonia, phenols, and oils. Some of them also contain mercaptains, benzo-a-pyrene, and other toxic organics. Since this industry generates considerable quantities of hazardous wastes, some small-quantity generators may dispose of hazardous wastes in on-site, land-based facilities.	Approximately 59% of the wastes are managed in on-site land application facilities. The remaining 41% are managed at off-site land-based disposal sites.
Pharmaceutical preparations	GOOD: The quantities and types of wastes from this industry are fairly well-described and the general waste management methods are known. (Year—1976)	LOW: The majority of these wastes are fermentation products and are biodegradable.	Approximately 85 to 90% of the wastes from the industry are managed in off-site, land-based disposal facilities.
Plastics and resins manufacturing	VERY GOOD: Detailed information is available on all data areas except the design features of the waste management facilities. (Year—1982)	HIGH: Many of the waste streams in this industry contain organic solvents and unreacted monomers, which are frequently toxic.	Approximately 68% of these wastes are treated in surface impoundments, 1% are landfilled, and 1.5% are managed in offsite, land-based disposal facilities.

Continued

Table 15 *Continued*

Industry	Data availability[a]	Relative levels of heavy metals or organics in wastes	Prevalent waste management methods
Primary iron and steel manufacturing and ferrous foundries	GOOD: The waste types and quantities generally are available, and the compositions of each waste are known. Management methods generally are known for each waste type. (Year—1983)	HIGH: Many of the wastes from this industry are low in pH and may release significant quantities of heavy metals.	Approximately 25% of these wastes are managed in on-site impoundments and landfills. Also, 65% of the wastes (mainly slag) are stored in waste piles prior to recycling.
Primary nonferrous metals manufacturing and nonferrous foundries	POOR: Good descriptions of the types of wastes produced by each sector, but not much analytical data. Good estimates on the quantities of each waste type, but almost no waste management data. (Year—1984)	HIGH: Several of the waste streams contain high levels of heavy metals.	No data.
Rubber and miscellaneous plastic products	POOR: Good data on quantities of wastes, but poor descriptions of waste characteristics and management methods. (Year—1975)	HIGH: Data are sketchy, but indicate possibly significant levels of elastomers, carbon black, plastic resins, plasticizers, and pigments.	At least some on-site landfilling and incineration, but data are almost nonexistent.

Soaps; other detergents; polishing, cleaning, and sanitation goods	POOR: Waste types poorly defined and quantity data are almost nonexistent. (Year—1974)	LOW: Most of these wastes are composed of packaging, lost products, salts, inerts. Some organics are generated from floor polishes (plasticizers) and pine oils (solvents).	Most of these wastes are expected to be sent off site because the industry is composed of a large number of small establishments.
Stone, clay, glass, and concrete products	POOR: Waste quantity data are available only for some waste types. Waste types are fairly well described, but lack analytical data. Management methods are poorly documented.	LOW: Most of the wastes produced are inert, earth-type materials. However, significant quantities of air pollution control sludges are generated, some of which may contain heavy metals.	No data; however, most wastes are expected to be managed on site due to generally low toxicity and high volumes.
Textile manufacturing	POOR: Waste types are fairly well described, but there are virtually no analytical data and no data on waste quantities and management methods.	LOW: Waste descriptions indicate low organics and heavy metals, but there are virtually no analytical data to confirm this assumption.	No data.

Continued

Table 15 *Continued*

Industry	Data availability[a]	Relative levels of heavy metals or organics in wastes	Prevalent waste management methods
Transportation Equipment	POOR: There are no data in the literature pertaining to non-hazardous waste generation and management within this industry.	HIGH: Wastes are expected to be similar in quantity and composition to those generated within SIC 34 and 35. Since this industry generates considerable quantities of hazardous wastes, some small-quantity generators may dispose of hazardous wastes in on site, land-based facilities.	No data.
Water treatment	POOR: Waste types are fairly well described and an overall estimate on waste quantities was available; however, there were no data on waste management methods.	LOW: These wastes are composed mainly of alum and lime, but may contain some heavy metals.	No data.

[a] Data areas pursued in this study included detailed analyses on each type of waste generated by each industry, the amount of each type of waste, the types and numbers of on-site, land-based disposal methods used by each industry, the general design of these facilities, and the amounts of each waste type managed in each different type of facility. The year for which most data were found is given in parentheses.